EE·5

Solid Surface
Luminescence Analysis

MODERN MONOGRAPHS IN ANALYTICAL CHEMISTRY

Series Editor

GEORGE G. GUILBAULT

Department of Chemistry
University of New Orleans
New Orleans, Louisiana

1. Solid Surface Luminescence Analysis: Theory, Instrumentation, Applications, *Robert J. Hurtubise*

    Additional Volumes in Preparation

Solid Surface Luminescence Analysis

THEORY, INSTRUMENTATION, APPLICATIONS

Robert J. Hurtubise
The University of Wyoming
Laramie, Wyoming

MARCEL DEKKER, INC.   New York and Basel

Library of Congress Cataloging in Publication Data

Hurtubise, Robert J.   [date]
   Solid surface luminescence analysis.

   (Modern monographs in analytical chemistry; 1)
   Includes indexes.
   1. Luminescence spectroscopy.   2. Surface chemistry.
I. Title.   II. Series.
QD96.L85H87   543'.0858   80-27916
ISBN 0-8247-1265-X

MARCEL DEKKER, INC.
270 Madison Avenue, New York, New York   10016

Current printing (last digit):
10 9 8 7 6 5 4 3 2 1

PRINTED IN THE UNITED STATES OF AMERICA

To Paula, Tim, Dave, and Suzanne

Foreword

Tremendous advances have been made in the field of analytical chemistry in the past ten years--new techniques, improved instrumentation, sophisticated automation, and new methods of data processing and handling. In an attempt to keep analytical chemists abreast of these rapidly changing developments the idea of the current series was conceived.

Modern Monographs in Analytical Chemistry will be a series of books, each devoted to a special treatise of a particular subject or field. Subjects to be presented will cover a large area from lasers to automation, piezoelectric crystals to fluorescence assays in biology and medicine. Each book will give an informative introduction to the field for the nonexpert, present a thorough description of the types of research currently being conducted, and conclude with a discussion of the future of this area. It will appeal to the expert as well as the novice. It is anticipated that monographs will appear in the developing areas of current research.

George G. Guilbault

Preface

Solid surface luminescence analysis has been applied to the trace
analysis of elements and compounds, but mainly to the analysis of
organic compounds. The versatility of solid surface luminescence
analysis is illustrated by the variety of samples that have been
analyzed. Several commercial instruments are in use and many modified or laboratory-constructed instruments have appeared for solid
surface luminescence analysis. The theoretical understanding of
luminescence scattered from solid surfaces has been developed par-
tially and remains to be expanded. With the advent of room temperature phosphorescence from solid surfaces and high-performance thin-
layer chromatography, new dimensions have been added to solid surface
luminescence analysis. Some exciting developments should be seen
in these two areas in the future.

Detailed consideration of solid surface luminescence analysis
is given in this book in the areas of theory, instrumentation, and
applications. To the author's knowledge no other book has appeared
on the subject that specifically treats solid surface luminescence
analysis. Two chapters are devoted to instrumentation, one to theoretical aspects of luminescence scattered from solid surfaces, another
to interactions responsible for room temperature phosphorescence,
one to some analytical procedural considerations, seven to analytical
applications, and one to future trends.

The beginner, the practicing analyst, and the researcher should all find this book useful in their work. Also, the nonanalytical chemist who wishes to use luminescence analysis or who already uses luminescence analysis will find the book helpful. In addition, the book should be useful as a textbook to teach the theoretical, instrumental, and applied aspects of solid surface luminescence analysis. Basic theory on luminescence was not presented because of the ready availability of several fine textbooks that consider luminescence theory.

Appreciation is extended to Dr. J. D. Winefordner for the use of some prepublication material on room temperature phosphorescence. Also, I am indebted to Agnes Kennington and Karen Singer, who typed the original manuscript, and to Sharon Breitweiser, who typed the final version.

Robert J. Hurtubise

Contents

Contents

1

Introduction

Solid surface luminescence analysis involves the measurement of
fluorescence or phosphorescence of components adsorbed on solid
materials. A variety of solid materials has been employed in
analysis, such as silica gel, aluminum oxide, filter paper, silicone rubber, sodium acetate, potassium bromide, sucrose, and cellulose. Both inorganic and organic species can be analyzed using
solid surfaces. Many of the arguments advanced for sensitivity
and selectivity for solution luminescence analysis are applicable
directly to solid surface luminescence analysis (1-3). Luminescence spectrometric methods are among the most sensitive methods
available for trace analysis. Because many excellent books have
already been written on solution luminescence analysis which include a discussion of the theoretical distinction between fluorescence and phosphorescence (1-3), those aspects will not be
considered here.

One fundamental difference between solution luminescence
and solid surface luminescence is that in solid surface luminescence the luminescent components are usually adsorbed on small
particles of the solid material, and this results in the scattering of both source and luminescent radiation. The radiation,
whether it is source radiation or luminescent radiation, can be
reflected from the solid material or transmitted through the solid

1

material. Of course for transmitted radiation, the experimental
system has to allow for transmission and measurement of the radiation such as with a plastic-backed thin-layer chromatoplate.
Wendlandt and Hecht (4) discussed the differences between specular
reflection and diffuse reflection. These concepts are important
in solid surface luminescence analysis. *Specular reflection* or
mirror reflection occurs from a very smooth surface and is described by the Fresnel equations. *Diffuse reflection* of exciting
radiation results through penetration of the incident radiation
into the interior of the solid material and multiple scattering
occurs at the boundaries of individual particles. Ideal diffuse
reflection is defined by the condition that the angular distribution of the reflected radiation is independent of the angle
of incidence (5). Several theories have been developed to describe diffuse reflection. Kortum (5) has emphasized that specular reflection and diffuse reflection are two important limiting
cases, and all possible variations are found in practice between
these two extremes. However, with solid surface luminescence
analysis diffuse luminescence is measured normally. Generally,
in solid surface luminescence analysis, a fraction of the sample
penetrates into the solid matrix, and the sample luminescence
is excited at the surface and within the solid matrix. The excited luminescence is scattered diffusely. As will be discussed
in Chapter 4, the theoretical aspects of scattered luminescence
in a solid matrix are not fully developed. In this book,
reflected luminescence refers to diffusely reflected luminescence,
and it appears at the same side as the excitation source. Any
specular reflection is considered unimportant. *Transmitted
luminescence* refers to diffusely transmitted luminescence, or
luminescence that appears at the unexcited surface. Wendlandt
and Hecht (4) and Kortum (5) have treated in detail the diffuse
reflectance of external source radiation used in absorption
measurements. They have given a brief treatment of luminescence
generated within a solid medium (4,5). In Chapter 4, a discus-

sion of recent theoretical equations useful in solid surface luminescence analysis will be considered.

A variety of commercial and laboratory-constructed instruments are used to measure solid surface luminescence. Commercial instruments and accessories are available for simple routine analysis, and more sophisticated models are used for fundamental research. The laboratory-constructed instruments are such that the design principles employed by the researchers can be used for further developing instrumentation in solid surface luminescence analysis.

Hundreds of applications have appeared in the literature since the introduction of commercial instrumentation for solid surface luminescence analysis. Several commercial instruments became available about 1968. The applications have appeared in such areas as environmental research, forensic science, pesticide analysis, food analysis, pharmaceutical process control, biochemistry, medicine, and clinical chemistry. The many examples to be considered in this book illustrate the simplicity, accuracy, sensitivity, and versatility of solid surface luminescence analysis. Recently room temperature phosphorescence from compounds adsorbed on solid surfaces has been shown to be feasible analytically (6-17). This approach will expand substantially the versatility of solid surface luminescence analysis, and many applications which use room temperature phosphorescence will appear in the future.

REFERENCES

1. G. G. Guilbault, *Practical Fluorescence*, Marcel Dekker, Inc., New York, 1973.

2. J. D. Winefordner, S. G. Schulman, and T. C. O'Haver, *Luminescence Spectrometry in Analytical Chemistry*, Wiley-Interscience, New York, 1972.

3. D. M. Hercules, ed., *Fluorescence and Phosphorescence Analysis*, Interscience Publishers, New York, 1966.

4. W. W. Wendlandt and H. G. Hecht, *Reflectance Spectroscopy*, Interscience Publishers, New York, 1966.

5. G. Kortum, *Reflectance Spectroscopy*, Springer-Verlag, New York, 1969.

6. M. H. Roth, *J. Chromatogr., 30*, 276 (1967).

7. E. M. Schulman and R. T. Parker, *J. Phys. Chem., 81*, 1932 (1977).

8. R. A. Paynter, S. L. Wellons, and J. D. Winefordner, *Anal. Chem., 46*, 736 (1974).

9. T. Vo-Dinh, G. L. Walden, and J. D. Winefordner, *Anal. Chem., 49*, 1126 (1977).

10. R. M. A. von Wandruszka and R. J. Hurtubise, *Anal. Chem., 49*, 2164 (1977).

11. C. D. Ford and R. J. Hurtubise, *Anal. Chem., 51*, 659 (1979).

12. C. G. de Lima and E. M. de M. Nicola, *Anal. Chem., 50*, 1658 (1978).

13. M. L. Meyers and P. G. Seybold, *Anal. Chem., 51*, 1609 (1979).

14. T. Vo-Dinh and J. R. Hooyman, *Anal. Chem., 51*, 1915 (1979).

15. G. L. Walden and J. D. Winefordner, *Appl. Spectrosc., 33*, 166 (1979).

16. L. J. Cline Love, M. Skrilec, and J. G. Habarta, *Anal. Chem., 52*, 754 (1980).

17. C. D. Ford and R. J. Hurtubise, *Anal. Chem., 52*, 656 (1980).

2

Commercial Instruments

Instruments for quantitative thin-layer chromatography (TLC) have
been available since about 1968. The commercial instruments can
be employed for quantitative and qualitative analysis of a variety
of luminescence components adsorbed on several solid surfaces,
such as silica gel, alumina, filter paper, gels, potassium bromide, silicone rubber, and sodium acetate. Reviews on commercial
instruments have appeared that describe self-contained units and
motorized thin-film scanners as attachments to spectrofluorome-
ters (1-7). Only a few commercial instruments will be considered
here.

The Schoeffel SD 3000 is a spectrodensitometer that can be
used in several different spectral modes. The instrument is shown
in Fig. 2.1, and an optical diagram is given in Fig. 2.2. Either
a 200-W xenon-mercury lamp or a 150-W xenon lamp can be used as
the exciting source. The solid surface such as a TLC plate with
fluorescent components on it is positioned on the stage of the spectrodensitometer and quartz optics focus the source radiation
onto the solid surface, after dispersion by a prism monochromator.
The area illuminated can be controlled in both width and length.
Figure 2.2 shows the instrument in several modes of operation.
When fluorescence or phosphorescence is measured, the single-
beam mode is employed because the only reference available is

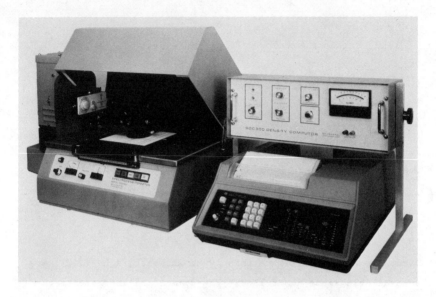

FIG. 2.1 Schoeffel SD 3000 spectrodensitometer. (Reproduced with permission from Kratos, Inc., Schoeffel Instrument Division.)

the nonfluorescent background of the solid surface. Transmitted or reflected luminescence can be measured with detectors either from underneath a transparent surface or at a 45° angle to the incident source radiation for measurement from a solid surface. Interference wedge monochromators covering the range from 400 to 650 nm graduated into 10-nm divisions can be employed in the reflection and transmission emission modes, or else appropriate filters can be used. To obtain luminescence emission spectra, a grating emission monochromator that can detect radiation from 180 to 720 nm can be attached to measure reflected luminescence. Also, the SD 3000 can be employed in the fluorescence quenching mode, and a paper strip, cylindrical gel, and gel slab scanning attachments are available.

The Kontes Glass Company has available an inexpensive densitometer that is useful for routine quantitative fluorescence analysis in the visible region for components separated on TLC plates (Fig. 2.3). The instrument can be used in the fluorescence and

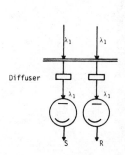

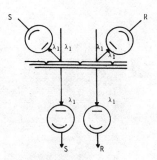

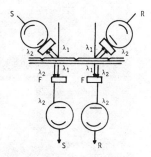

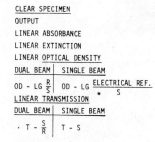

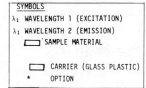

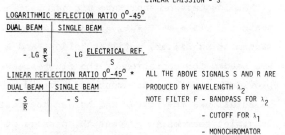

CLEAR SPECIMEN
OUTPUT
LINEAR ABSORBANCE
LINEAR EXTINCTION
LINEAR OPTICAL DENSITY

DUAL BEAM	SINGLE BEAM
OD - LG $\frac{R}{S}$	OD - LG $\frac{\text{ELECTRICAL REF.}}{S}$ *

LINEAR TRANSMISSION

DUAL BEAM	SINGLE BEAM
$\cdot$ T - $\frac{S}{R}$	T - S

SYMBOLS
λ_1 WAVELENGTH 1 (EXCITATION)
λ_1 WAVELENGTH 2 (EMISSION)
⬚ SAMPLE MATERIAL

⬚ CARRIER (GLASS PLASTIC)
* OPTION

OPAQUE SPECIMEN
OUTPUT
LINEAR ABSORBANCE
LINEAR EXTINCTION
LINEAR OPTICAL DENSITY

DUAL BEAM	SINGLE BEAM
OD - LG $\frac{R}{S}$	OD - LG $\frac{\text{ELECTRICAL REF.}}{S}$

LINEAR TRANSMISSION RATIO*

DUAL BEAM	SINGLE BEAM
T - $\frac{S}{R}$	T - S

LOGARITHMIC REFLECTION RATIO 0^0-45^0

DUAL BEAM	SINGLE BEAM
- LG $\frac{R}{S}$	- LG $\frac{\text{ELECTRICAL REF.}}{S}$

LINEAR REFLECTION RATIO 0^0-45^0 *

DUAL BEAM	SINGLE BEAM
- $\frac{S}{R}$	- S

FLUORESCENCE
FLUORESCENCE QUENCHING
LINEAR ABSORBANCE

DUAL BEAM	SINGLE BEAM
OD - LG $\frac{R}{S}$	OD - LG $\frac{\text{ELECTRICAL REF.}}{S}$

LINEAR FLUORESCENCE QUENCHING RATIO

DUAL BEAM	SINGLE BEAM
- $\frac{S}{R}$	- S

FLUORESCENCE MEASUREMENTS
LINEAR EMISSION - S

ALL THE ABOVE SIGNALS S AND R ARE
PRODUCED BY WAVELENGTH λ_2
NOTE FILTER F - BANDPASS FOR λ_2
                                    - CUTOFF FOR λ_1
                                    - MONOCHROMATOR

FIG. 2.2 Schoeffel SD 3000 spectrodensitometer modes of operation.
(Reproduced with permission from Kratos, Inc., Schoeffel Instrument
Division.)

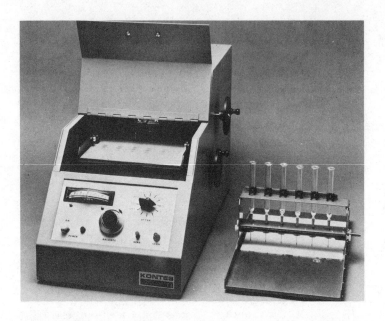

FIG. 2.3 Kontes densitometer. (Reproduced with permission from Kontes Glass Company.)

fluorescence quenching modes, or diffuse reflectance and transmission modes for absorption analysis. In the fluorescence mode
either longwave or shortwave ultraviolet lamps, which are underneath the plate carriage, are used for excitation. An ultraviolet
transmitting filter, a filter holder, and a quartz cover plate
are needed for operation of the fluorescence mode. They are not
provided with the instrument and have to be purchased separately.
In actual use the filter and holder are positioned over the ultraviolet sources to prevent the passage of visible light. The
developed chromatoplate is placed on top of the quartz cover plate
with the adsorbent side face down. This assembly is placed on
the plate carriage and the fluorescent components are excited
in single-beam operation with either shortwave or longwave ultraviolet radiation. The visible fluorescence of the components
passes through the chromatoplate's glass or plastic backing and
is transmitted by fiber optics to a photodetector. A choice of

several acetate filters is provided that may be placed in front
of the photodetector and pass fluorescent radiation of interest.
With TLC spots less than 5 mm in diameter, a "reference" head
is employed, and for bigger spots a "read" head is employed for
efficient gathering of fluorescence. In the fluorescence quenching mode dual-beam operation is possible which allows the fluorescence of the fluorescent phosphor in the TLC absorbent to be compared with fluorescence remaining after partial quenching of the
fluorescent phosphor by the separated components.

    Listed in Table 2.1 are several manufacturers that sell
densitometers or thin-film scanners as attachments for spectro-
fluorometers.

TABLE 2.1   Some Suppliers of Densitometers or Scanning Attachments

Supplier	Address
American Instrument Company	8030 Georgia Avenue Silver Spring, Maryland 20910
Carl Zeiss, Inc.	444 Fifth Avenue New York, New York 10018
Farrand Optical Co., Inc.	117 Wall Street Valhalla, New York 10595
Gelman Instrument Company	600 South Wagner Road Ann Arbor, Michigan 48106
Kontes Glass Company	Vineland, New Jersey 08360
Perkin-Elmer	Norwalk, Connecticut 06856
Schoeffel Instrument Corporation	24 Booker Street Westwood, New Jersey 07675
Shimadzu Scientific Instruments, Inc.	Oakland Ridge Industrial Center 9147-H Red Branch Road Columbia, Maryland 21045
Transidyne General Corporation	903 Airport Drive Ann Arbor, Michigan 48104
V-tech Corporation	16229 W. Ryerson Road P. O. Box 183 New Berlin, Wisconsin 53131

REFERENCES

1. R. J. Hurtubise, P. F. Lott, and J. R. Dias, *J. Chromatogr. Sci.*, *11*, 476 (1973).

2. P. F. Lott, J. R. Dias, and R. J. Hurtubise, *J. Chromatogr. Sci.*, *14*, 488 (1976).

3. P. F. Lott, J. R. Dias, and S. C. Slahck, *J. Chromatogr. Sci.*, *16*, 571 (1978).

4. R. W. Frei and J. D. MacNeil, *Diffuse Reflectance Spectroscopy in Environmental Problem-Solving*, CRC Press, Cleveland, Ohio, 1973, pp. 31-79.

5. M. S. Lefar and A. D. Lewis, *Anal. Chem.*, *42*, 79A (1970).

6. G. G. Guilbault, *Photochem. Photobiol.*, *25*, 403 (1977).

7. J. F. Lawrence and R. W. Frei, *Chemical Derivatization in Liquid Chromatography*, Vol. 7, Elsevier Scientific Publishing Co., New York, 1976, pp. 48-60.

3

Modified Instruments, Accessories, and Experimental Techniques

Several researchers have designed and built their own instruments
and accessories for a variety of measurements from solid surfaces.
In this chapter, instruments and accessories that can be employed
for the measurement of luminescence reflected or transmitted from
solid surfaces will be considered. Instruments have been built
that function only in the absorption or transmission modes, but
if redesigned, they could be used in the luminescence mode. Some
of these instruments and accessories will also be discussed. The
optical theoretical principles behind these instruments such
as the theory of Kubelka and Munk are reserved for Chapter 4.
This chapter is primarily concerned with instrument design, accessories, and innovative experimental techniques.

INSTRUMENTS FOR ABSORPTION MEASUREMENTS

Goldman and Goodall (1) modified a Chromoscan densitometer (manufactured by Joyce Loebl & Co. Ltd., Gateshead, England) to use
for quantitative analysis in the absorption mode of components
separated on thin-layer chromatoplates. The essential modification involved a scanning apparatus of their design that fitted
into the sample compartment of the Chromoscan densitometer.
The scanning apparatus gave a sawtooth motion to the chromatoplate

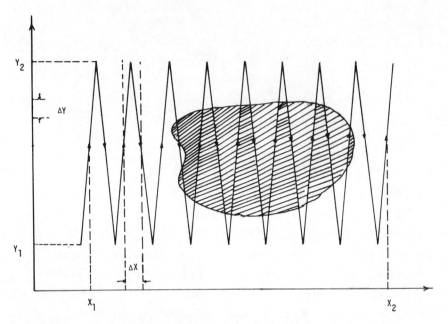

FIG. 3.1 Diagrammatic representation of the path of the light spot (→—) over a chromatographic zone (hatched area). (Reprinted with permission from Ref. 1.)

as illustrated in Fig. 3.1. The approach is sometimes called the flying spot technique. Because the absorbance of the separated components varies both in length and breadth, the sawtooth motion of the scanning device should allow more accurate measurement of absorbance. However, with this approach a large number of data points is generated, necessitating the use of a computer for data processing. With some redesigning Goldman and Goodall's scanning apparatus possibly could be used with instruments for luminescence measurements for improved accuracy. However, as will be discussed in Chapter 4 the flying spot technique theoretically does not offer great advantages in luminescence measurements.

Goldman and Goodall (2) built an instrument that allowed the measurement of transmittance of silica gel chromatoplates in the ultraviolet region down to 240 nm. The instrument calculated and recorded on computer tape the ultraviolet transmittance at 0.5 × 0.5 mm intervals in the two dimensions of the chromatoplate. Silica gel absorbs radiation in the region 200-280 nm, which is an important consideration in absorption analysis and luminescence analysis. Goldman and Goodall (2) calculated that the absorption by silica gel decreased the transmittance of a typical silica gel chromatoplate to 0.001 of that normally encountered above 280 nm. Thus the system they designed required a powerful ultraviolet source and a highly sensitive detecting system. The instrument is shown diagrammatically in Fig. 3.2. The components were in modular form with the exception of the scanning mechanism, lamp housing, and chopper. The scanning mechanism drove the chromatoplate in 0.5-mm steps in a square-wave motion of adjustable amplitude. The time interval between steps was synchronized with the computer tape output. The light path and optical components are shown in Fig. 3.2. The source was a water-cooled deuterium lamp. After the source radiation passed through a 1213-Hz chopper, the radiation was focused by a condenser onto the chromatoplate and then passed through the Suprasil support plate. A portion of radiation transmitted through the chromatoplate was focused onto a monochromator. The entrance and exit slits of the monochromator were masked so that no more than a 0.5×0.5 mm region of the chromatoplate was observed at any instant. Radiation from the monochromator fell onto a 10-mm diameter end-window cathode of a photomultiplier. The ac output from the photomultiplier was taken into a three-module low-noise amplifier/phase shifter/phase-sensitive detector system where the signal was gated with a reference signal to provide a low-noise dc output. The reference signal was obtained from a silicon photovoltaic cell which was illuminated by a small tungsten filament lamp. The signal voltage was converted to a three-digit number and punched onto computer tape at up to 100 characters per second. The data on the computer tape was proportional to transmittance and was converted to absolute transmittance by measurement of a reference standard of known transmittance. Dilute

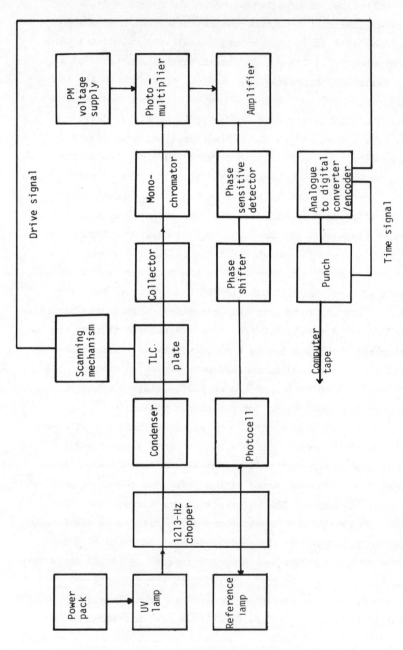

FIG. 3.2 Diagrammatic representation of the instrument showing the interrelation of the functional components. (Reprinted with permission from Ref. 2.)

India ink was used as a reference standard. The computer program employed by Goldman and Goodall for data processing carried out two iterative procedures at each data point, aligned the data, determined the length of each zone, and interpolated the background transmittance in the zone. The computing time per zone was approximately 45 sec on a KDF 9 computer. Although Goldman and Goodall's instrument was not designed to measure luminescence, some of the design principles could be employed for luminescence instrumentation. Sophisticated instruments of this type for measurement of luminescent signals have not appeared commercially. Goodall (3) modified the instrument discussed previously so that transmission could be presented continuously as a negative logarithm on a flatbed recorder in the ultraviolet or visible regions. With their earlier instrument the signals were digitized and recorded on paper tape, and sometimes there were delays in processing the tapes. Thus Goodall designed an instrument that allowed both encoded digital tape and instantaneous graphical recordings. This permitted rapid review of results to decide if more accurate computer processing was required. A block diagram of the instrument is shown in Fig. 3.3.

AN INSTRUMENT FOR FLUORESCENCE MEASUREMENTS

Pollak and Boulton (4-11), Boulton and Pollak (12,13), and Pollak (14-19) have discussed extensively instrument theory and design for quantitative absorption analysis of components on solid surfaces such as thin-layer chromatoplates. Pollak and Boulton (10) and Pollak (17,20-22) considered theoretically the performance of photometric methods for the quantitative evaluation of thin-layer chromatoplates using fluorescence. Their concepts will be considered in Chapter 4. An instrument designed by Pollak and Boulton and Gietz that can be employed in the reflectance and transmittance absorption modes and the fluorescence mode (23, 24) will be considered here. Even though the instrument was

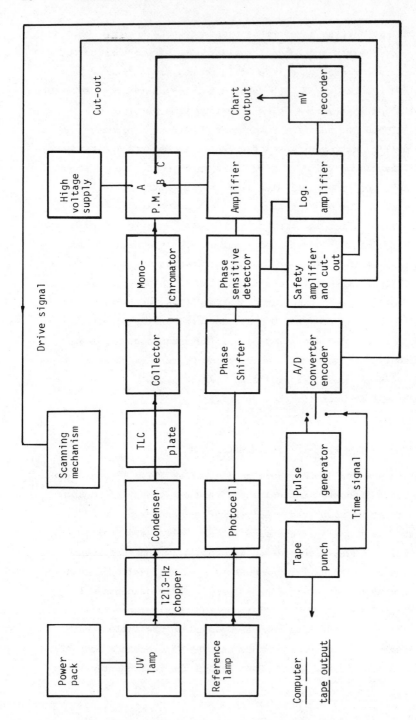

FIG. 3.3 Diagram of the revised instrument which now offers either encoded digital tape or analog output on a chart. (Reprinted with permission from Ref. 3.)

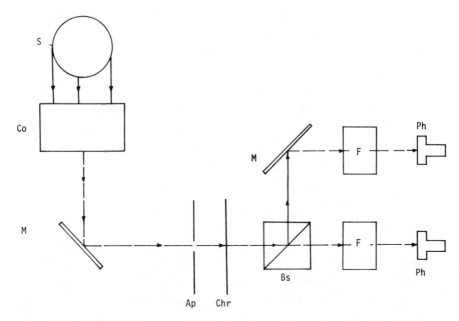

FIG. 3.4 Schematic representation of the optical path of a double-beam scanning device: S, light source; Co, collimator; M, mirror; Ap, aperture; Chr, chromatogram; Bs, beam splitter; F, optical filter or monochromator; Ph, photodetector. (Reprinted with permission from Ref. 23.)

designed mainly for absorption measurements, useful data were obtained in the fluorescence mode. This illustrates the similarity of instrument design for absorption and luminescence measurements. Pollak and Boulton were interested in designing an instrument that could be employed in all types of photometric determinations which are used in thin-layer chromatography and related techniques.

They considered several features for optimum performance of the instrument. One of these was a double-beam design that avoided any spatial or time separation of the beams before the beams interacted with the chromatoplate (Fig. 3.4). Also they adopted the principle of flying spot scanning to compensate for any nonhomogeneous distribution of sample within a measured spot zone. The two beam signals were linearized in terms of concentration by logarithm forming for transmittance and simple inversion for reflectance measurements. The purpose of logarithm forming was twofold. First, optical noise was decreased substantially and the signal was almost independent of the output of the light source. Second, a signal was obtained which was nearly a linear function of the concentration. This latter aspect is very important for a flying spot system to operate efficiently. For solid surface luminescence analysis, linearization is not required (10, 17). This is an important consideration in designing instruments for solid surface luminescence analysis and will be discussed theoretically in Chapter 4. Pollak (20) pointed out that the double-beam method is advantageous for fluorescence measurements but the combination of the two beams to minimize optical noise is less straightforward compared to transmission or reflection measurements. Also, the theoretical principles for luminescence measurements are still in the developmental stages and need to be further clarified for instrument design. With Pollak and Boulton's instrument, fluorescence was measured in the single-beam mode. The fluorescence signal was passed by a buffer amplifier to an analog integrator. The integrator was basically a high-gain dc amplifier with a resistive-capacitive feedback network between the output and the input. The integrator was started at the beginning of each scan line by a trigger pulse derived from the position of a scanning drum. The integrator integrated the signal over the duration of one scan line. At the end of a scan line, the integrator was reset. The value obtained was stored in a sample-and-hold circuit until the end of the following integration period. From the sample-and-hold circuit the value was passed by a buffer amplifier to the output. A staircase output wave was produced, and the height of each stair corresponded to the integral of the fluorescence signal over the last scan line. A standard analog recorder was used to display the output. For a smooth output curve, a smoothing filter could be inserted into the output circuit ahead of the recorder.

Boulton, Gietz, and Pollak (24) reported data on the stability and reproducibility of their new instrument. Only a discussion of the fluorescence data will be given here. The fluorophore dansyl ethylamine which was separated by paper chromatography was used to evaluate the performance of the instrument. Also the instrument was employed only in the fluorescence transmission mode. A straight-line relationship was obtained for the fluorescence of dansyl ethylamine when log area (arbitrary units) versus log sample (nanograms) was plotted. The reproducibility of the instrument at both high and low amounts was determined by scanning the same chromatogram usually 16 times. At high amounts (500 ng) the deviation from the mean was ±0.7%, and at low amounts (10 ng) the deviation from the mean was ±12%. The authors emphasize that reproducibility would be improved if the scanning signal were converted to digital form and assessed by a computer. The evaluation of the overall analytical procedure was accomplished by preparing several chromatograms and subjecting them to the analytical procedure. Because of the additional experimental manipulation the standard deviation increased. Despite the increase the error was quite acceptable analytically. The instrument was capable of handling thin-layer absorbents on supports such as glass, polyethylene, and aluminum, provided their width was 5 cm or less.

RECOMMENDED IMPROVEMENTS FOR INSTRUMENTS

Treiber (25) discussed recently the development and practical utilization of a linear spectrodensitometer. He modified a commercial instrument and showed that it had high sensitivity, reliability, versatility, and short analysis times for biologic applications and for monitoring of organic reactions during synthetic work. Even though Treiber did not investigate solid surface

luminescence analysis, some of his conclusions are applicable
to this area. He emphasized that there are no commercial instruments available that can be used for quantitative work in a convenient way without modification or addition of various parts.
The same general conclusions apply to instrumentation for solid
surface luminescence analysis, but as will be shown in the chapters on applications, commercial instruments can be employed
successfully for numerous applications. Below are some of Triber's
conclusions that can be used in designing improved instruments
for solid surface luminescence analysis.

1. The instrument should locate components on a solid sur-
    face in an automatic manner.

2. The instrument should allow scanning in a two-dimensional
    fashion.

3. The instrument should accommodate several solid surfaces
    such as thin-layer chromatoplates and scan them succes-
    sively in an automatic manner.

4. The instrument should have digital outputs that give
    the location of the components, area under curves for
    luminescence scans, and the excitation and emission wave-
    lengths.

INSTRUMENTATION FOR SOLID SURFACE FLUORESCENCE OF SPECIES
IN BIOLOGIC SYSTEMS

Guilbault and co-workers (26-30) have designed equipment and developed methods for the assay of enzymes, substrates, activators,
and inhibitors. Several of their applications will be described
later in this book. In this section the instrumentation, accessories, and preparation of solid surfaces will be discussed.

Generally, an Aminco filter instrument was placed on its
side (Fig. 3.5) and a cell holder was adapted to accept a metal
slide on which a silicone rubber pad was placed that contained

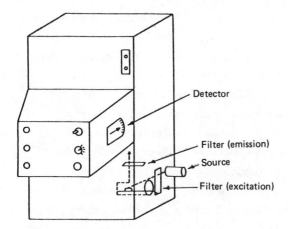

FIG. 3.5 Fluorometer placed on end for horizontal pad studies.
(Reprinted with permission from Ref. 30.)

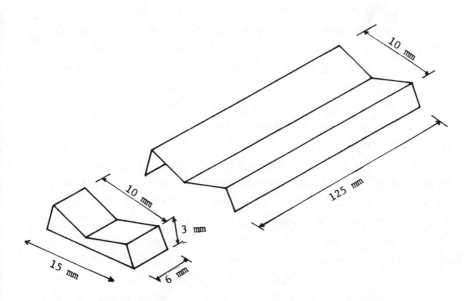

FIG. 3.6 Silicone rubber pads of 6-mm width (lower left) are cut
from a 125-mm strip (upper right). (Reprinted with permission from
Ref. 30.)

the necessary reagents for an assay. The change in fluorescence with time was measured and related to the concentration of the substance to be determined. A more detailed discussion of this successful and unique approach follows (30).

The silicone rubber pads were prepared by pressing uncured rubber between a glass plate and a stainless steel mold (Fig. 3.6). Both the glass plate and the stainless steel mold were lined with a piece of glassine paper. The surfaces contacted with the silicone rubber were prelubricated with silicone stopcock grease. The silicone rubber in the mold was cured for 2 days at room temperature. The cured strips were washed briefly with concentrated KOH solution and then washed with H_2O and dried at 80°C for 1 hr. Individual pads 6 mm wide were cut from the strip, and about 20 individual pads could be obtained from one strip. The reagent pads were simple to prepare and hundreds could be easily manufactured at one time. The pads were stable for months and even longer when stored under specified conditions. For solid surface fluorescence analyses, essentially all the reagents are present on the pad and this minimizes the need for the cumbersome, time-consuming preparation of reagents. For situations where sample size is limited, the pad method is very advantageous, since only 3-25 µl of sample is required for an assay (30).

To form the reactant film on the pad, the reagents and buffer in an appropriate solvent were deposited on the silicone pad so the solution spread evenly over the pad. The solution was then evaporated off by vacuum or lyophilization. The reagent could also be applied to the pad in a polymeric film such as polyacrylamide, or a stabilizer could be added if an enzyme was in the reagent. Substrate, enzyme, or coenzyme could be deposited on the pad in film form, depending on whether the substance to be determined was an enzyme or a substrate. The pads could be stored conveniently before use in a dark, cold place or in a refrigerator or desiccator. Guilbault and co-workers (30) found that the most accurate results were obtained if gray silicone rubber pads were used. Also, the

silicone pads could retain a reactant film on their surfaces for
an indefinite time and allow the direct detection of fluorescence
from their surfaces when an appropriate reagent was applied onto
the first reactant film.

For all fluorometric measurements, an Aminco filter fluorometer
was employed. It was set on its side to prevent any reagents from
falling off the surface of the pad (Fig. 3.5). Two wooden blocks
were used to support the fluorometer to prevent electrical noise
and to make it convenient to change the primary filter when needed.
The results were displayed with a linear recorder.

Two kinds of cells and cell holders were designed by Guilbault
and co-workers. The first kind was designed by Guilbault and Vaughan
(27) to study reactions at room temperature. It was modified for
temperature control so reactions could be studied above 30°C. Only
the second type will be discussed, because this device can be used
for room temperature or higher temperature studies. The cell holder
was an Aminco cuvet adapter (Cat. No. J4-7330) with water circulating around it to maintain constant temperature. Black binders were
placed on both sides of the two entrance and exit slits of the cell
holder. This allowed smaller slits approximately two-thirds the
length of the pad, which restricted the radiation that entered and
left the cell cavity. A cylindrical aluminum rod with a slot, about
twice the length of the pad, positioned toward the end of the rod,
was used to construct the cell (Fig. 3.7). The cell was painted
black to avoid scattered radiation. The cell was designed so the
pad with its contents received the full beam of incident radiation.
Figure 3.8 shows the pad inside the cell holder and its relation
to the incident radiation (30).

INSTRUMENTATION FOR ROOM TEMPERATURE PHOSPHORESCENCE

An exciting new area of analytical research is room temperature
phosphorescence (RTP). Roth (31) first suggested the analytical
use of RTP from his observations that several organic compounds

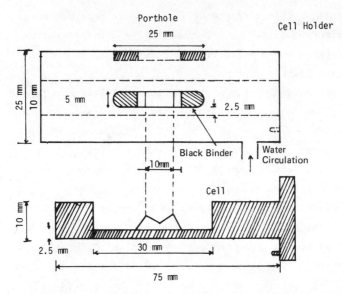

FIG. 3.7 The cell (lower drawing) is constructed of a cylindrical aluminum rod with a slot, approximately twice the length of the pad, located toward the end of an Aminco cuvet adapter with water circulating around it to maintain a constant temperature. (Reprinted with permission from Ref. 30.)

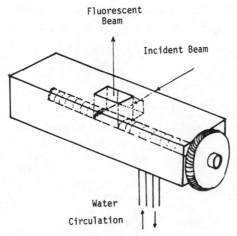

FIG. 3.8 Pictured here are the pad inside the cell holder, and the relationship of the pad to the incident beam. (Reprinted with permission from Ref. 30.)

adsorbed on cellulose gave RTP. Lloyd and Miller (32) recently commented on the first observations of RTP. Schulman and Walling (33,34)
independently measured RTP from several compounds. Paynter, Wellons,
and Winefordner (35) analytically developed and applied the phenomena
observed by Walling and Schulman using filter paper as a solid surface. Von Wandruszka and Hurtubise (36,37) first showed the analytical application of sodium acetate as a solid surface for RTP.
Seybold and White (38) and Vo-Dinh, Yen, and Winefordner (39) discussed the heavy-atom effect on room temperature phosphorimetry.
Applications and other aspects of RTP will be considered in Chapter 7.

Paynter, Wellons, and Winefordner (35), in developing their
new method of analysis based on RTP, used an Aminco-Bowman spectrophotofluorometer with a phosphoroscope attachment without a quartz
dewar flask. They employed filter paper as a solid surface and carried out a thorough study of the phosphorescence background intensity
of various filter papers. They found that Eaton Dikeman 613 filter
paper gave the lowest phosphorescence background. The filter paper
was cut into 1/4-in. circles using a paper hole puncher. The researchers suspended the circles of paper vertically in clothesline
fashion with alligator clips. A volume of 5 μl of sample was deposited on each filter paper circle by allowing the sample solution
to drain slowly from the tip of the syringe when contacted with
the filter paper. Drying of the sample was necessary because moist
samples showed considerable decrease in phosphorescence signals.
After testing many methods of drying, infrared lamp heating was found
to be gentle and very effective (40). The dried sample was placed
in a long spindlelike holder that fitted directly into a standard
Aminco phosphoroscope accessory instead of the normal cap and cylinder
used for the dewar-flask assembly. To keep the sample compartment
dry, dry air was passed through the system. After drying, the phosphorescence signal was measured.

Vo-Dinh, Walden, and Winefordner (41) designed an automatic
phosphorimetric instrument for RTP with a continuous filter paper
device. A block diagram of the instrumental system is shown in

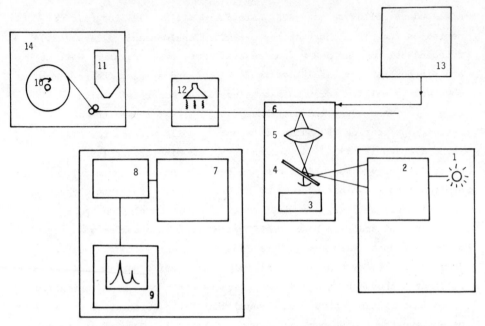

FIG. 3.9 Schematic diagram of an AutoAnalyzer continuous filter
with the room temperature phosphorescence detection system. (1) Light
source, (2) excitation monochromator, (3) rotation motor-phosphoro-
scope, (4) reflecting surface, (5) optics, (6) filter paper, (7) emission monochromator, (8) detection unit, (9) recorder, (10) filter
paper roll, (11) spotting syringe, (12) drying IR lamp, (13) dry
air supply, (14) AutoAnalyzer continuous filter. (Reprinted with
permission from T. Vo-Dinh, G. L. Walden, and J. D. Winefordner,
Anal. Chem., 49, 1126 (1977). Copyright by the American Chemical
Society.)

Fig. 3.9. The detection unit of the instrumental system was an

Aminco-Bowman spectrophotofluorometer. The sample compartment of

the unit was modified and equipped with a laboratory-constructed

rotating mirror assembly for detection of phosphorescence. The

rotating mirror and operating principle is shown schematically in

Fig. 3.10. A diagonally cut section of an aluminum cylindrical

rod was employed for the mirror and reflecting surface. The sur-

face was well polished for good reflection of ultraviolet and visible

radiation. Excitation radiation from the excitation monochromator

was reflected by the reflecting surface onto the surface of the

filter paper (Fig. 3.10). The filter paper moved horizontally across

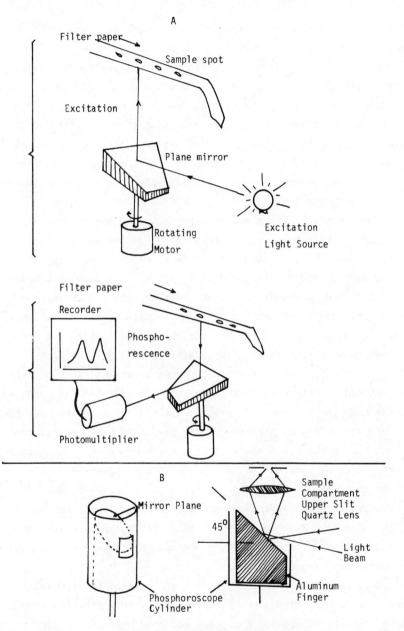

FIG. 3.10 (A) Principle of phosphorimetric excitation and detection with the rotating mirror phosphorimeter. (B) Design of the rotating mirror assembly. (Reprinted with permission from T. Vo-Dinh, G. L. Walden, and J. D. Winefordner, *Anal. Chem., 49*, 1126 (1977). Copyright by the American Chemical Society.)

a slit located at the top of the sample compartment. The reflection plane moved into the emission path as the cylindrical reflecting surface rotated, and the phosphorescence emitted by the sample
was reflected back into the detection system. The excitation radiation was not "seen" by the detector during the excitation period.
Also, scattered radiation was decreased greatly by inserting the
reflecting cylindrical surface into an Aminco-Keirs phosphoroscope
attachment (Fig. 3.10). At the upper part of the sample compartment an aperture allowed a continuous warm-air flow over the filter
paper during an experiment.

A Technicon Continuous Filter paper roll was employed as a
solid support and was drawn into the drying chamber and the cell
compartment (Fig. 3.9). Samples were delivered drop by drop with
a hypodermic syringe onto the moving filter paper by spotting manually 3 µl of sample solution. After spotting, the filter paper
was fed into the drying chamber where the sample remained for about
2 min. The paper was then passed over the sample compartment of
the Aminco-Bowman spectrophotofluorometer (Fig. 3.9). Vo-Dinh,
Walden, and Winefordner (41) found that two important experimental
variables had an influence on the phosphorescence intensity measured
with their system. These were predrying the samples before measurement and the continuous flushing of dry gas through the sample compartment during the measurement. For the compounds they investigated
optimal predrying time was between 5 and 10 min. Several series
of 10 to 15 identical samples of different materials were measured.
The relative standard deviations for most series of measurements
was less than 5%. With the instrumental system developed and the
compounds investigated, the limit of detection was in the nanogram
and subnanogram ranges.

Later, Yen-Bower and Winefordner (42) designed an improved
experimental system with a new filter paper guide which allowed continuous sampling of organic phosphors adsorbed on filter paper.
Phosphorimetric time resolution was achieved by an analog switch
phosphoroscope (43). The filter paper guide was placed in the sample

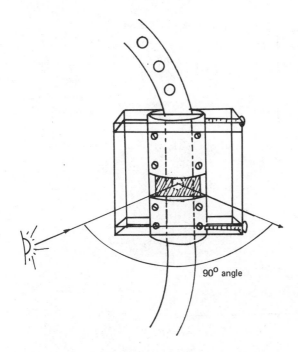

90° angle

FIG. 3.11   Schematic diagram of assembled filter paper guide for
the continuous sampling system.   (Reprinted with permission from
Ref. 42.)

compartment of an Aminco-Bowman spectrophotofluorometer in place
of the Aminco-Keirs rotating can phosphoroscope. Generally the
analytical data indicated that the new system would be very useful
in areas where several samples are handled routinely. The assembled
filter paper guide is shown in Fig. 3.11. Before the filter paper
passed into the sample compartment, the paper entered a drier, and
a Teflon roller was placed in the drier so that the filter paper
moved smoothly. Several holes in the drier allowed warm (60°C),
dry nitrogen to flush the drying area. Filter paper was fed to the
sample compartment from a filter paper roll and was removed from
the sample compartment by means of a Technicon Autoanalyzer Continuous
Filter.

Walden and Winefordner (44) have made a comparison of ellipsoidal
and parabolic mirror systems in fluorometry and room temperature

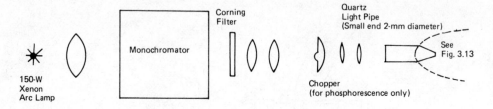

FIG. 3.12 Diagram of the excitation optical system. (Reprinted
with permission from Ref. 44.)

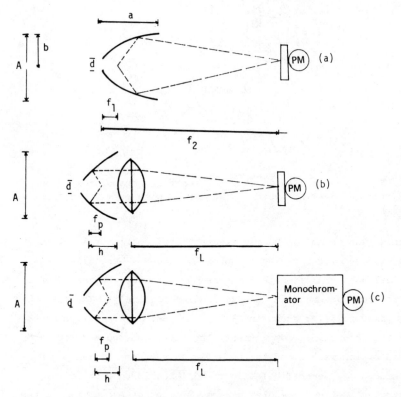

FIG. 3.13 Diagram of the detector system used incorporating (a) the
ellipsoidal mirror with filter, (b) the parabolic mirror with filter,
and (c) the parabolic mirror with monochromator. For the ellipsoid:
f_1 = 1 in., f_2 = 7.56 in., d = 0.25 in., A = 5.5 in., a = 4.28 in.,
b = 2.75 in. For the paraboloid: f_p = 0.75 in., f_L = 12 in., d = 0.63
in., A = 4.85 in., h = 1.95 in. (Reprinted with permission from
Ref. 44.)

phosphorimetry. With the mirrors they were able to increase the
collection efficiency of sample luminescence for small volume samples.
Walden and Winefordner emphasized that one problem with most commercial luminescence instruments is that only a small fraction of the
total 4π sr of the emitted luminescence is collected and measured.
An instrument with f/4 collection optics collects approximately
0.015π sr of the total emitted luminescence. Figure 3.12 shows
the excitation system employed. The excitation light was focused
onto the end of a quartz light pipe that was tapered on the opposite
end to a diameter of approximately 2 mm. This had the effect of
approximately focusing the excitation light on the sample when the
sample was placed near the end of the light pipe. For the ellipsoidal
and parabolic mirror systems investigated, a small hole was drilled
through the center back of the mirror and the light pipe shoved into
the hole just shy of the focal point (Fig. 3.13). The luminescence
emission wavelengths were selected by either filters or a monochromator, and an RCA 1P28 photomultiplier tube was used as a detector.
For the RTP studies, the excitation radiation was chopped with a
laboratory-constructed phosphoroscope. The RTP samples were spotted on filter paper circles, dried, and placed in the sample compartment at the end of a metal rod (41). Table 3.1 gives a comparison
of the RTP limits of detection (signal equal to 3 times the blank
luminescence standard deviation) for several compounds. Walden

TABLE 3.1 Room Temperature Phosphorimetry Absolute Limits of
Detection (ng)

Compound	Aminco SPR	Parabolic mirror system monochromator	Ellipsoidal mirror system filter
Phenanthrene	0.9	2	0.02
Carbazole	0.7	0.2	1
Tryptophan	0.2	1	8
1,2,5,6-Dibenzanthracene	8	0.6	0.09
8-Amino-1-naphthol-3,6- disulfonic acid	0.5	0.2	0.09

Reprinted with permission from Ref. 44.

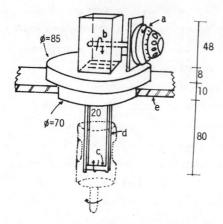

FIG. 3.14 Design of the paper chromatography accessory. (a) Helipot
dial; (b) brass traction pin; (c) lower brass pin; (d) rotating
can phosphoroscope; (e) sample compartment wall (all measurements
in mm). (Reprinted with permission from C. G. de Lima and E. de M.
Nicola, *Anal. Chem.*, *50*, 1658 (1978). Copyright by the American
Chemical Society.)

and Winefordner did not give results for the parabolic mirror-filter
system because the limits of detection were similar to the ellipsoidal
mirror-filter system. The authors commented that with RTP on filter
paper, emission from the front surface is considerably greater than
emission levels from the back surface. Thus, the effective solid
angle of emission is $<2\pi$ sr rather than $\sim 3\pi$ sr. Walden and Winefordner
have shown that high collection efficiency mirrors can be very useful
for analysis of samples of small area. They plan to use laser excitation so extremely small volumes can be studied and a true point
source can be approached. The major problem with the systems is
that the sample has a finite volume. This means that a true point
source is not achieved in the mirror.

De Lima and de M. Nicola (45) employed a laboratory-assembled
single-monochromator spectrophosphorimeter with a rotating can phosphoroscope to measure RTP of 1,8-naphthyridine derivatives adsorbed
on filter paper. They constructed a multiple sample paper chromatography accessory for measurement of RTP, shown in Fig. 3.14. A
rubber band was placed around two brass pins (band c in Fig. 3.14)

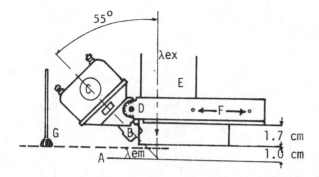

FIG. 3.15 Modified reflection mode assembly. (A) RTP solid support;
(B) detector lens; (C) photomultiplier tube housing; (D) photomultiplier tube housing angle adjustment screw; (E) optical nose of
spectrodensitometer; (F) slide adjust screws for varying distance
from exciting radiation to photomultiplier tube housing; (G) rotating
disk. (Reprinted with permission from C. D. Ford and R. J. Hurtubise,
Anal. Chem., 51, 659 (1979). Copyright by the American Chemical
Society.)

which were machine grooved in order to provide traction. The bank
was driven manually by a Helipot dial which also served to locate
the spot. A dried-chromatography Whatman No. 1 paper strip with
samples on it was attached to the band with tape. An air blower
was used for initial drying, and a vacuum oven was used for final
drying.

Ford and Hurtubise (46) designed a phosphoroscope and reflection
mode assembly for use with a Schoeffel SD 3000 spectrodensitometer
to measure RTP of nitrogen heterocycles adsorbed on silica gel and
filter paper. The reflection mode assembly is shown in Fig. 3.15.
The assembly allowed the distance from the source exit to the photomultiplier tube to be varied by means of an adjustable slide. Also,
the angle of the photomultiplier tube housing could be adjusted
to maximize the reflected RTP striking the lens photodetector system.
The preceding modifications were necessary to accommodate a rotating
disk phosphoroscope. The phosphoroscope assembly consisted of a
variable speed dc motor (0-12 V) and a rotating disk phosphoroscope.
The motor was powered by a Railine model 370N transformer. The assembly was designed for use with the modified reflection mode assembly

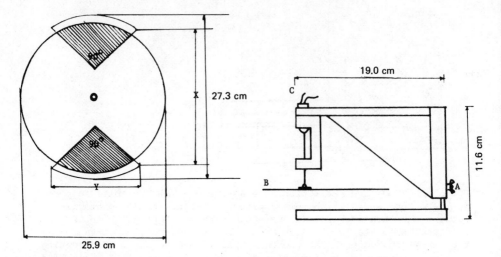

FIG. 3.16 The disk (a) and mount (b) of the phosphoroscope assembly.
(A) Wingnut for disk height adjustment; (B) rotating disk; (C) motor.
(Reprinted with permission from C. D. Ford and R. J. Hurtubise,
Anal. Chem., 51, 659 (1979). Copyright by the American Chemical
Society.)

of the spectrodensitometer (Fig. 3.15). The phosphoroscope (Fig.
3.16) was constructed of thin sheet aluminum and was painted flat
black to minimize scattered radiation. The dc motor was mounted
in an aluminum frame which allowed the phosphoroscope to be raised
or lowered via an adjustable slide (Fig. 3.16). In region X (Fig.
3.16) the compound on the solid surface was excited by the source
radiation; however, the lens which collects the reflected luminescence
and focuses it onto the photomultiplier tube was positioned so no
luminescence was detected. As the phosphoroscope rotated region
Y (Fig. 3.16) over the adsorbed compound, the compound was no longer
excited and the detector system detected any delayed luminescence.
The reflection mode assembly and phosphoroscope have general applicability with the Schoeffel SD 3000 spectrodensitometer in measuring
RTP from solid surfaces.

     Schulman and Parker (47) used an Aminco-Keirs spectrophosphori-
meter with a rotating can phosphoroscope and a sample holder similar
to the one described by Paynter, Wellons, and Winefordner (35) for

holding paper circles with samples deposited on the circles. Seybold
and White (38), von Wandruszka and Hurtubise (36,37), and Ford and
Hurtubise (48) employed a Perkin-Elmer MPF-2A fluorescence spectrophotometer with a rotating can assembly for some of their recent
work. Seybold and White used filter paper as a solid surface, von
Wandruszka and Hurtubise used sodium acetate, and Ford and Hurtubise
used silica gel. Jakovljevic (49) investigated filter paper (Whatman
No. 42) impregnated with lead tetraacetate or thallous acetate for
the RTP of several organic compounds. He constructed a sample holder
in which filter paper strips were inserted, and used an Aminco-Bowman
spectrophotofluorometer with a cold finger phosphoroscope to obtain
phosphorescence excitation and emission spectra.

INSTRUMENTATION FOR LOW-TEMPERATURE PHOSPHORESCENCE

Gifford, et al. (50) constructed and evaluated a thin-layer, single-
disk, multislot phosphorimeter for the direct measurement of phosphorescence from separated components on thin-layer chromatograms
near liquid nitrogen temperature. Their phosphorimeter can also
be used to measure RTP. In a later paper the authors described
the construction and evaluation of an improved thin-layer phosphorimeter (51). The thin-layer phosphorimeter was designed to fit the
sample compartment of a Baird-Atomic Fluoricord spectrofluorometer
and is illustrated in Fig. 3.17. A thin-layer chromatoplate was
affixed with elastic bands to the outside of a hollow copper sample
drum which could be filled with liquid nitrogen. The separated
samples were on aluminum-backed silica gel chromatoplates or filter
paper. Cellulose acetate and aluminum oxide thin layers were also
employed (50). The bottom of the drum was lipped to allow positioning
on a turntable in the holder compartment. The rate of rotation of
the turntable was controlled by a variable output transformer and
a scanning rate of 3-40 cm min^{-1} was provided. Two silica windows
were fitted in the outer cylinder of the sample holder compartment.
The windows allowed incident radiation to reach the sample on the
solid surface at a 45° angle to the normal, and the emitted radiation

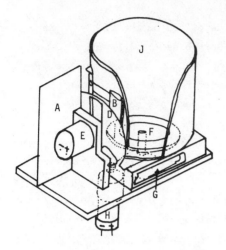

FIG. 3.17 Sectional diagram of the thin-layer phosphorimeter attachment, showing the (A) light baffle, (B,C,) slits, (D) disk phosphoroscope, (E) phosphoroscope motor, (F) turntable, (G) intermediate gear, (H) turntable driving motor, and (J) outer cylinder. The sample drum is omitted for the sake of clarity. (Reprinted with permission from J. N. Miller, D. L. Phillipps, D. T. Burns, and J. W. Bridges, *Anal. Chem.*, *50*, 613 (1978). Copyright by the American Chemical Society.)

that reached the detector was observed at a 45° angle to the normal. The outer surfaces of the silica windows were continuously swept by a stream of dry oxygen-free nitrogen which prevented the formation of ice and minimized luminescence quenching by oxygen. A single-disk phosphoroscope assembly was used. The disk was 65 mm in diameter and had three equally spaced slots cut in it and was driven up to 10,500 rpm by a 12-V electric motor. When a single 5-μl sample of benzophenone on silica gel was scanned repeatedly using the thin-layer phosphorimeter, a relative standard deviation of 4% was obtained for measured peak intensities.

When samples were separated by TLC before scanning, reproducible results were obtained only when the chromatoplate was sprayed with a solvent such as ethanol and an internal standard was employed. With these conditions the relative standard deviation for several components was about 8%. Spraying the developed chromatoplate was found to enhance the phosphorescence of separated components

considerably (51). The authors (50) have discussed the design of
their single-disk, multislot phosphoroscope in detail. O'Haver
and Winefordner (52) have described the relation between phosphorescence intensity for rotating cylinder and Becquerel, twin-disk phos-
phoroscopes. Gifford et al. (50) considered the ability of their
phosphoroscope to resolve short- and long-lived phosphorescence and
related this to four time periods. These were (1) the cycle time
t_c, the time for one complete cycle of excitation and observation,
(2) the exposure time t_e, the time during which the sample is excited
or observed, (3) the transit time t_t, the time for the rotating
aperture to open fully, (4) the delay time t_d, the time between
the end of one excitation period and the beginning of the next observation period. They compared disks of different slot dimensions
and showed under what design conditions the phosphorescence intensity
became constant.

LASERS IN SOLID SURFACE LUMINESCENCE ANALYSIS

There has been practically no use of lasers in solid surface luminescence analysis. Certainly much potential exists for the application of lasers in this area. Berman and Zare (53) used time-resolved
and wavelength-resolved laser-induced fluorescence for the analysis
of aflatoxins on TLC chromatoplates. Their experimental setup is
shown schematically in Fig. 3.18. They employed a pulsed nitrogen
laser (337.1 nm) as a source and a RCA 7265 photomultiplier with appropriate apertures and wavelength filters for detection of fluorescence. As the authors stated, a boxcar integrator was the heart
of the instrumental setup which controlled when the photomultiplier
output was interrogated. The trigger generator controlled the firing
of the nitrogen laser and sent a start impulse to the boxcar integrator. The boxcar integrator had a preset delay before opening
an electronic gate to sample the signal from the photomultiplier.
Also, the boxcar integrator had a preset "window" so the photomultiplier signal was sampled for a fixed time. The laser was fired
repetitively, and each pulse had a duration of about 14 nsec with

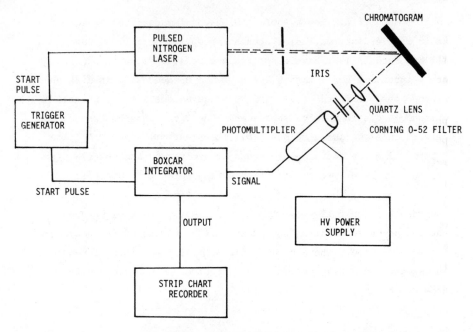

FIG. 3.18 Schematic of an experimental setup. (Reprinted with permission from M. R. Berman and R. N. Zare, *Anal. Chem.*, *47*, 1200
(1975). Copyright by the American Chemical Society.)

a power of approximately 100 kW. The photomultiplier signals were
collected and averaged by the boxcar integrator, and the output
of the integrator drove a strip-chart recorder which displayed the
fluorescence signal as a function of time and thus chromatoplate
displacement. Berman and Zare emphasized that the technique of
laser fluorescence analysis would have several advantages over fluorescence detection by standard techniques. These are (1) the laser
radiation source is much brighter than conventional sources; (2) the
laser radiation source is coherent and may be focused on a spot
more readily; (3) the effect of phosphorescence and scattered laser
radiation is minimized because gated detection electronics are employed; (4) the use of time-resolved and wavelength-resolved detection
should allow the analysis of mixtures of fluorescent species without
separation of the components. The authors commented that much work
remains to demonstrate the applicability of laser fluorescence analysis; however, their preliminary results are encouraging.

VIDICON RAPID SCANNING DETECTOR FOR CHEMILUMINESCENCE TLC DETECTION

Curtis and Seitz (54) used a silicon Vidicon detector system to detect chemiluminescence from components separated by TLC. Generally the Vidicon detection head is a two-dimensional array of photodiodes. When photons strike a photodiode they cause a buildup of charge. Using an electron beam which completely scans the array once every 32 msec the charge is discharged and measured. The detector surface in their system was divided into 500 separate channels in the horizontal direction. The intensity striking each channel was measured by a multichannel analyzer and displayed on a cathode-ray tube. In their work, optical dispersion as with a grating was not used. The emission from the chromatoplate was focused onto the Vidicon detector surface. With this approach, different detector channels corresponded to different positions on the chromatoplate. The cathode-ray tube output of intensity versus channel number was then a record of emission intensity versus plate position along one axis. A silicon Vidicon detector offers several advantages for measuring chemiluminescence from chromatoplates. Sensitivity is greater because the detector simultaneously measures emission from all areas of the chromatoplates along one dimension rather than focusing at one particular spot at a given time. The authors emphasized that with conventional fluorescence, it would be necessary to design a system for illuminating the chromatoplate evenly as well as for keeping source radiation from reaching the detector. With chemiluminescence, the luminescence is generated chemically on the chromatoplate and there is no need for an external source of radiation.

INDUCED FLUORESCENCE BY GASEOUS ELECTRICAL DISCHARGE

Shanfield, Hsu, and Martin (55) found that a wide variety of organic compounds adsorbed on silica gel chromatoplates gave intense fluorescence after a special electrical treatment. The treatment involved placing a piece of the plate with adsorbed compound on it into an electrical discharge tube which was evacuated to approximately 0.2

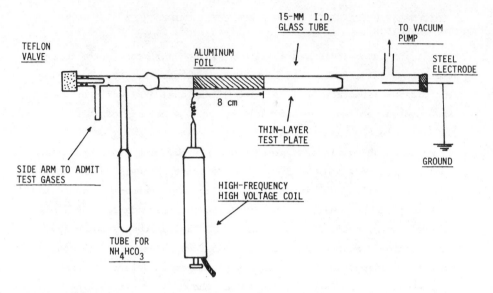

FIG. 3.19 Schematic diagram of electrical discharge chamber for inducing fluorescence in organic compounds spotted or separated on silica gel thin-layer chromatoplates. (Reprinted with permission from Ref. 55.)

torr, and discharging the system for a period of 5 sec to 5 min. After treatment in the tube, the plate was heated on a hot plate at about 130°C. The fluorescence was observed under a 365-nm ultraviolet lamp until it appeared to reach a maximum intensity.

A schematic diagram of the electrical discharge chamber is shown in Fig. 3.19. The discharge was produced by a Tesla coil-type high-frequency generator which was rated at a maximum of 20,000 V, 0.5 MHz. The coil is the type used routinely to detect leaks in glass vacuum systems. The high-frequency output was connected to the plasma chamber by aluminum foil wrapped externally around a section of a glass tube (Fig. 3.19). A steel-rod electrode provided the second electrode and was connected to ground. An intense glow filled the tube from within the foil covered region to the steel electrode when in operation. As shown in Fig. 3.19, a side chamber was provided for ammonium hydrogen carbonate, and a provision was made to admit gases into the chamber. Segura and Gotto (56) induced fluorescence in several organic compounds by exposing them to vapors

of ammonium hydrogen carbonate. Shanfield, Hsu, and Martin (55)
obtained observable fluorescence from all the compounds tested in
the atmosphere resulting from ammonium hydrogen carbonate. With
a gaseous environment of nitrogen or hydrogen, the fluorescence
produced was about equivalent to an ammonium hydrogen carbonate
environment. Some of the compounds investigated were cholesterol,
lindane glucose, phenobarbitol, and phenol. Signals were obtained
at the nanogram and microgram levels for these compounds. Much
more work has to be done with electrically induced fluorescence
to extend its analytical applicability. However, from the variety
of organic compounds shown to give fluorescence up to now, it appears
that in the future the approach should have wide application for
organic trace analysis.

REFERENCES

1. J. Goldman and R. R. Goodall, *J. Chromatogr.*, *40*, 345 (1969).

2. J. Goldman and R. R. Goodall, *J. Chromatogr.*, *47*, 386 (1970).

3. R. R. Goodall, *J. Chromatogr.*, *103*, 265 (1975).

4. V. Pollak and A. A. Boulton, *J. Chromatogr.*, *45*, 200 (1969).

5. V. Pollak and A. A. Boulton, *J. Chromatogr.*, *46*, 247 (1970).

6. V. Pollak and A. A. Boulton, *J. Chromatogr.*, *50*, 19 (1970).

7. V. Pollak and A. A. Boulton, *J. Chromatogr.*, *50*, 30 (1970).

8. V. Pollak and A. A. Boulton, *J. Chromatogr.*, *50*, 39 (1970).

9. V. Pollak and A. A. Boulton, *J. Chromatogr.*, *63*, 87 (1971).

10. V. Pollak and A. A. Boulton, *J. Chromatogr.*, *72*, 231 (1972).

11. V. Pollak and A. A. Boulton, *J. Chromatogr.*, *76*, 393 (1973).

12. A. A. Boulton and V. Pollak, *J. Chromatogr.*, *45*, 189 (1969).

13. A. A. Boulton and V. Pollak, *J. Chromatogr.*, *63*, 75 (1971).

14. V. Pollak, *J. Chromatogr.*, *63*, 145 (1971).

15. V. Pollak, *J. Chromatogr.*, *77*, 245 (1973).

16. V. Pollak, *Trans. I.E.E.E.*, *17*, 287 (1970).

17. V. Pollak, *Opt. Acta*, *21*, 51 (1974).

18. V. Pollak, *J. Chromatogr.*, *105*, 279 (1975).

19. V. Pollak, *Opt. Acta*, *23*, 25 (1976).

20. V. Pollak, *J. Chromatogr., 133*, 49 (1977).

21. V. Pollak, *J. Chromatogr., 133*, 195 (1977).

22. V. Pollak, *J. Chromatogr., 133*, 199 (1977).

23. V. Pollak and A. A. Boulton, *J. Chromatogr., 115*, 335 (1975).

24. A. A. Boulton, W. Gietz, and V. Pollak, *J. Chromatogr., 115*, 349 (1975).

25. L. R. Treiber, *J. Chromatogr., 123*, 23 (1976).

26. G. G. Guilbault and A. Vaughan, *Anal. Lett., 3*, 1 (1970).

27. G. G. Guilbault and A. Vaughan, *Anal. Chim. Acta, 55*, 107 (1971).

28. H. K. Y. Lau and G. G. Guilbault *Enzym. Tech. Dig. 3*, 164 (1974).

29. R. L. Zimmerman and G. G. Guilbault, *Anal. Chim. Acta, 58*, 75 (1972).

30. G. G. Guilbault, "Fluorescence Analysis on Solid-Surfaces," in *Essays in Memory of Anders Ringdom*, E. Wänninen, ed., Pergamon Press, New York, 1977, pp. 435-452.

31. M. Roth, *J. Chromatogr., 30*, 276 (1967).

32. J. B. F. Lloyd and J. N. Miller, *Talanta, 26*, 180 (1979).

33. E. M. Schulman and C. Walling, *Science, 178*, 53 (1972).

34. E. M. Schulman and C. Walling, *J. Phys. Chem., 77*, 902 (1973).

35. R. A. Paynter, S. L. Wellons, and J. D. Winefordner, *Anal. Chem., 46*, 736 (1974).

36. R. M. A. von Wandruszka and R. J. Hurtubise, *Anal. Chem., 48*, 1784 (1976).

37. R. M. A. von Wandruszka and R. J. Hurtubise, *Anal. Chem., 49*, 2164 (1977).

38. P. G. Seybold and W. White, *Anal. Chem., 47*, 1199 (1975).

39. T. Vo-Dinh, E. L. Yen, and J. D. Winefordner, *Anal. Chem., 48*, 1186 (1976).

40. S. L. Wellons, R. A. Paynter, and J. D. Winefordner, *Spectrochim. Acta*, Pt. A, *30*, 2133 (1974).

41. T. Vo-Dinh, G. L. Walden, and J. D. Winefordner, *Anal. Chem., 49*, 1126 (1977).

42. E. L. Yen-Bower and J. D. Winefordner, *Appl. Spectrosc., 33*, 9 (1979).

43. E. L. Yen, G. D. Boutilier, and J. D. Winefordner, *Can. J. Spectrosc., 22*, 120 (1977).

44. G. L. Walden and J. D. Winefordner, *Appl. Spectrosc., 33*, 166 (1979).

45. C. G. de Lima and E. de M. Nicola, *Anal. Chem., 50*, 1658 (1978).

46. C. D. Ford and R. J. Hurtubise, *Anal. Chem.*, *51*, 659 (1979).

47. E. M. Schulman and R. T. Parker, *J. Phys. Chem.*, *81*, 1932 (1977).

48. C. D. Ford and R. J. Hurtubise, *Anal. Chem.*, *50*, 610 (1978).

49. I. M. Jakovljevic, *Anal. Chem.*, *49*, 2048 (1977).

50. L. A. Gifford, J. N. Miller, D. T. Burns, and J. W. Bridges, *J. Chromatogr.*, *103*, 15 (1975).

51. J. N. Miller, D. L. Phillipps, D. T. Burns, and J. W. Bridges, *Anal. Chem.*, *50*, 613 (1978).

52. T. C. O'Haver and J. D. Winefordner, *Anal. Chem.*, *38*, 602 (1966).

53. M. R. Berman and R. N. Zare, *Anal. Chem.*, *47*, 1200 (1975).

54. T. G. Curtis and W. R. Seitz, *J. Chromatogr.*, *134*, 513 (1977).

55. H. Shanfield, F. Hsu, and A. J. P. Martin, *J. Chromatogr.*, *126*, 457 (1976).

56. R. Segura and A. M. Gotto, *J. Chromatogr.*, *99*, 643 (1974).

4

Theoretical Aspects of
Solid Surface Luminescence

Much work remains to be done on the theoretical characteristics
of solid surface luminescence phenomena as they relate to chemical
analysis. Almost no experimental data have been obtained to substantiate recent theoretical equations that describe the variation of
luminescence intensity from compounds adsorbed on solid surfaces.
What surface characteristics are best for room temperature phosphorescence of compounds has not been substantiated fully. Some headway
has been made on theoretical instrumental parameters for solid surface luminescence analysis. These and other aspects remain to be
investigated experimentally and theoretically. In this chapter,
theoretical equations that describe the relationship between the
characteristics of the solid medium and fluorescence intensity will
be considered. In addition the theoretical instrumental principles
will be considered for solid surface luminescence analysis.

KUBELKA-MUNK THEORY

Recent theoretical equations that have appeared in the literature
have their basis with the Kubelka-Munk theory (1). This theory has
been applied successfully in reflectance spectroscopy. Certainly
the potential exists for applying other theories to solid surface
luminescence analysis, but this information has not appeared in the

literature. Kortum (2) and Wendlandt and Hecht (3) have discussed
theories and applications related to reflectance spectroscopy.
A detailed treatment of the Kubelka-Munk theory will not be given
because this has been done elsewhere (1-3). However, some of the
fundamental assumptions and aspects of the Kubelka-Munk theory will
be considered below.

It is important to realize that almost all the surfaces employed
in solid surface luminescence analysis are highly scattering media.
That is, the solid materials such as silica gel and alumina scatter
both exciting radiation and luminescence radiation extensively.
In addition the adsorbed compound absorbs a fraction of the exciting
radiation, and it is possible for the solid surface itself to absorb exciting radiation. The two phenomena, absorption and scattering, are the main considerations in the Kubelka-Munk theory.
Many authors have discussed the relationship of the Kubelka-Munk
theory to thin-layer densitometry as applied in the transmission-
absorption and reflectance-absorption modes (4-10). Their ideas
are important because they emphasize many important aspects of the
Kubelka-Munk theory in densitometry and several of the concepts
are applicable to solid surface luminescence analysis.

Goldman and Goodall (4), Goldman (11), and Pollak and Boulton
(12) discussed the assumptions of the Kubelka-Munk theory related
to solid surface absorption and fluorescence analysis. An expanded
treatment of these assumptions applied to reflected radiation can
be found in textbooks on reflectance spectroscopy (2,3). Kubelka
(1) considered the assumptions for both reflected and external source
radiation in scattering materials. Several of the important assumptions of the Kubelka-Munk theory applicable to solid surface luminescence are as follows:

1. Radiation within the medium, whether source or luminescence
    radiation, propagates only in the forward and backward
    directions perpendicular to the boundary surfaces of the
    medium. The boundary surfaces are assumed to be plane-
    parallel (12).

2. A point on the surface of the medium is considered as an omnidirectional point source of radiation (12); in other words, once the radiation reaches the surface the radiation is scattered in all directions.

3. The intensity of radiation radiated by a point on the surface in a given direction is proportional to the cosine of the angle between that direction and the surface plan of the medium (12). A more detailed discussion of this phenomenon is given by Kortum (13).

4. The parameters that characterize the optical response of the medium are k, absorption coefficient of exciting radiation; s, scattering coefficient of exciting radiation; S, scattering coefficient of fluorescent radiation; and X, thickness of the scattering medium (1,11,12).

5. The direction of exciting radiation is perpendicular to the surface (12).

6. The medium is assumed to be homogeneous between the boundary surfaces. This means that theoretically the absorption coefficient and scattering coefficients are not a function of the thickness of the scattering medium (12).

GOLDMAN'S EQUATIONS

Goldman and Goodall (4,14,15) described the densitometric application of the absorptiometry theory of Kubelka and Munk. For fluorescence densitometry Goldman (11) took into consideration both excitation and fluorescent radiation in scattering media and obtained two pairs of differential equations. One pair corresponded to absorption of exciting radiation in transmitted and reflected directions and the other pair corresponded to fluorescent radiation emitted in the transmitted and reflected directions. With further mathematical treatment of the two pairs of differential equations Goldman arrived at two general equations that were very complex and would need a computer to process data. One equation represented fluorescence transmitted through a thin-layer chromatoplate and the other equation fluorescence

reflected from the surface of a thin-layer chromatoplate. These
complex equations have little analytical utility and will not be
emphasized. Goldman's complex equations can be manipulated many
ways. Under conditions of very low-level fluorescence and relatively
large values for scattering coefficients for both exciting radiation
and fluorescent radiation, Goldman presented the following equations:

$$I^+/i_0\alpha = 1/3kX(1 - 7/30\ sXkX) \tag{1}$$

$$J^+/i_0\alpha = 2/3kX(1 - 4/30\ sXkX) \tag{2}$$

where I^+ = intensity of transmitted fluorescence, i_0 = intensity
of initial exciting radiation, α = portion of absorbed radiation
converted into fluorescence, k = absorption coefficient of exciting
radiation, s = scattering coefficient of exciting radiation, X =
thickness of the scattering medium, and J^+ = intensity of reflected
fluorescence.

In equations (1) and (2) the scattering coefficient of fluorescent radiation does not appear because of mathematical approximations. In Goldman's more complex equations, he assumed the scattering
coefficient of exiciting radiation was equal to the scattering coefficient of fluorescent radiation for graphs of both $I^+/i_0\alpha$ and $J^+/i_0\alpha$
versus kX. The term kX is proportional to the amount of absorbing
compound in the exciting beam (4). Equations (1) and (2) will be
discussed more fully later.

When Goldman derived his fundamental equations he assumed that
the solid surface did not absorb exciting radiation. Goodall (16)
showed that common thin-layer chromatographic adsorbents such as
silica gel and aluminum oxide absorb radiation in the approximate
spectral range 225-340 nm. With solid surfaces that absorb exciting
radiation, Goldman's equations have to be modified to take this into
consideration. Goldman (11) discussed this and showed theoretically
that both transmitted fluroescence and reflected fluorescence can
be linear with the amount of fluorescent material adsorbed on the
solid surface.

Goldman has stated several other theoretical conclusions and
some of these are considered below (11).

At small levels of fluorescer the most advantageous analytical
mode is direct measurement of reflected fluorescence if the solid
surface is nonabsorbing. If the solid surface absorbs and the fluorescer is at a low level, then direct measurement can be carried
out, but layer thickness must be considered because the reflected
fluorescence is inversely proportional to the thickness of the layer.
Goldman stated that the transmission mode can be employed effectively
in those cases where the medium absorbs exciting radiation by adding
a fixed amount of some inert fluorescent substance to the solid
medium. Goldman showed that the amount of fluorescent substance
is proportional to $\left(I^* - I_0^*\right)/\left(I_0^*\right)^2$ which is linear in I^*, where
I^* is the fluorescence intensity of fluorescent component of interest,
and I_0^* is the background fluorescence of the inert fluorescent substance. Direct measurement of transmitted fluorescence is possible
and the background fluorescence level can be used to correct the
sample readings for layer thickness variation with the expression
$\left(I^* - I_0^*\right)/\left(I_0^*\right)^2$. Goldman has also considered the fluorescence quenching mode but it will not be discussed here. Goldman gave no experi-
mental data to support his theoretical conclusions.

EXPERIMENTAL DATA

Hurtubise (17) obtained experimental data from thin-layer chromatoplates using fluoranthene as a model compound to offer experimental
support for the simplified Goldman equations (1) and (2). A Schoeffel
SD 3000 spectrodensitometer and aluminum oxide (Al_2O_3) and silica
gel (SiO_2) glass-backed chromatoplates with an n-hexane mobile phase
were employed in the experimental work.

Experimental values of sX were calculated with the following
equation, $T_0 = 1/(sX + 1)$, where T_0 is the transmission of the chromatoplate when kX is zero (2). T_0 was calculated from the equation
$A_0 = \log 1/T_0$, and A_0 was determined experimentally.

The kX values for fluoranthene were calculated by successive

approximations with a programmable calculator using the general
Kubelka-Munk equation for transmission in scattering media (2,4).
The equation is as follows:

$$T = \frac{b}{a\ \sinh(bsX) + b\ \cosh(bsX)}$$

where T is transmittance of a chromatoplate with an absorbing material in the scattering medium,

$$a = \frac{s + k}{s} = \frac{sX + kX}{sX} \qquad b = (a^2 - 1)^{\frac{1}{2}}$$

It would be ideal to express kX simply in terms of T and sX; however,
this is quite complicated theoretically. In Hurtubise's work T
was calculated from the equation A = log 1/T, where A is equal to
A_0 plus the experimental absorbance for fluoranthene. Finally,
kX was calculated by successive approximations. It was assumed
that the chromatographic material did not absorb exciting radiation.
This was reasonable because Goodall (16) showed that silica gel and
aluminum oxide absorb very weakly around 340 nm. The excitation
wavelength used in Hurtubise's work was 370 nm. For the situation
where chromatographic material would absorb exciting radiation,
the absorption would have to be taken into consideration.

Once sX and kX were obtained, $I^+/i_0\alpha$ were calculated from equations (1) and (2). With these values and the corresponding kX values,
theoretical calibration curves were plotted for reflected and transmitted fluorescence. Also, experimental fluorescence calibration
curves were obtained for fluoranthene in the transmission and reflection modes and several comparisons were made with the theoretical
curves.

One comparison was the microgram value at which the theoretical
curves and experimental curves approximately first changed slope.
Data were obtained for fluoranthene on developed and undeveloped
aluminum oxide and silica gel chromatoplates to see if chromatographic
conditions had any influence on the reflected or transmitted fluorescence. The data obtained are given in Table 4.1 and show fairly
good comparison between theoretical and experimental values for

TABLE 4.1 Comparison of Theoretical and Experimental Approximate Points of First Slope Change in Terms of Microgram of Fluoranthene

| | Developed Al_2O_3 | | Undeveloped Al_2O_3 | | Developed SiO_2 | | Undeveloped SiO_2 | |
Plate	Theor.	Exptl.	Plate	Theor.	Exptl.	Plate	Theor.	Exptl.	Plate	Theor.	Exptl.
Transmitted fluorescence											
1	0.12	0.17	3	0.15	0.17	5	0.16	0.18	7	0.15	0.18
2	0.12	0.18	4	0.11	0.15	6	0.16	0.20	8	0.18	0.18
Reflected fluorescence											
1	0.12	0.14	3	0.21	0.15	5	0.16	0.22	7	0.19	0.13
2	0.13	0.15	4	0.16	0.14	6	0.14	0.15	8	0.18	0.16

Reprinted with permission from Ref. 17.

both developed and undeveloped chromatoplates, whether the fluorescence was measured in the transmission or reflection modes. The
results indicated that Goldman's simplified equations can be used
to predict approximately when a calibration curve first changes slope
with acceptable accuracy. Goldman (11) predicted the range of linearity of reflected fluorescence should be approximately twice that
obtained by transmission in terms of kX. The data in Table 4.1
do not support this conclusion. The data show the range of linearity of reflected and transmitted fluorescence are relatively close.

Figure 4.1 compares a theoretical and an experimental curve
with their relative intensity values normalized. As seen in Fig.

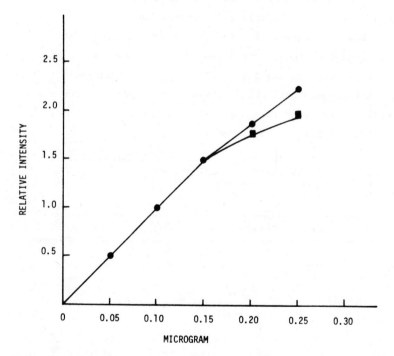

FIG. 4.1 Comparison of experimental (●) and theoretical (■) calibration curves from developed Al_2O_3 plate 1 in the fluorescence
transmission mode. (Reprinted with permission from R. J. Hurtubise,
Anal. Chem., *49*, 2160 (1977). Copyright by the American Chemical
Society.)

4.1, for an aluminum oxide chromatoplate the theoretical and experimental curves are superimposable to about 0.15 µg, and beyond that they diverge with the experimental curve showing larger relative intensity values per microgram of fluoranthene. One important contribution of the Goldman theory, considering the conditions employed in Hurtubise's work, is the capability of predicting approximately when the calibration curve first changes slope. However, further work is needed to explain the nonsuperimposability of the curves beyond the linear portion of the curves (17).

According to Goldman's simplified equations, the theoretical ratio of the slopes of a linear transmitted fluorescence calibration curve to a linear reflected fluorescence calibration curve is 0.50, namely, $(1/3)kX/(2/3)kX$. Table 4.2 compares the slopes of experimental calibration curves. The ratio of slopes for calibration curves obtained for the developed aluminum oxide and silica gel chromatoplates compares favorably with the theoretical ratio. Table 4.2 shows the experimental ratios for the developed aluminum oxide chromatoplates are lower than the theoretical ratio 0.50. The experimental ratios for the developed silica gel chromatoplates are

TABLE 4.2 Experimental Ratios of Slopes of Calibration Curves of Transmitted Fluorescence to Reflected Fluorescence

Plate	Al_2O_3 Developed	Al_2O_3 Undeveloped	SiO_2 Developed	SiO_2 Undeveloped
1	0.39			
2	0.36			
3		0.26		
4		0.51		
5			0.62	
6			0.58	
7				0.91
8				0.89

Reprinted with permission from Ref. 17.

higher than the theoretical ratio. This implies that the reflected
fluorescence from the aluminum oxide chromatoplates is relatively
greater than the reflected fluorescence from the silica gel chromatoplates. The experimental ratio for the undeveloped aluminum oxide
plate 3 (Table 4.2) was only 0.26. The low value suggests that the
fluoranthene did not penetrate the aluminum oxide chromatoplate
to any great extent because with a ratio of 0.26 the reflected fluorescence is substantially greater than the transmitted fluorescence.
The ratios for the undeveloped silica gel plates are close to 1.
This indicates that the fluorescent signals measured in the reflected
and transmitted modes were approximately the same. It was found
that with the undeveloped silica gel chromatoplates, the fluorescence
of fluoranthene at a given amount was much greater than its fluorescence on the developed silica gel chromatoplates. This suggested
the fluorescence quantum efficiency of fluoranthene was enhanced
considerably on undeveloped silica gel chromatoplates by an adsorption mechanism, or other mechanisms, different from the one for
fluoranthene when it was developed chromatographically. It appeared
the fluorescence intensity of fluoranthene was so great and the layer
thickness of the chromatoplates was such that signals in the reflection or transmission modes did not show a substantial difference
in fluorescence intensity. The data suggested that the fluorescence
intensity in the reflection and transmission modes is independent
over a range of layer thicknesses for strongly fluorescent components.
Goldman (11) did not consider these aspects in the development of
his theoretical equations. The simplified equations apparently
cannot be used to predict the ratio of relative fluroescence intensities in the transmitted and reflected modes for highly fluorescent
components over a certain range of layer thicknesses.

Hurtubise (17) presented experimental data that showed the
7/30 sXkX and 4/30 sXkX products for SiO_2 chromatoplates are smaller
than the corresponding products for Al_2O_3 chromatoplates. These
products appear in equations (1) and (2) and are important theoretically in determining the points at which fluorescence transmission

and reflection calibration curves become nonlinear. The experimental results indicated generally that the linear range for silica gel chromatoplates in terms of microgram (or kX) was greater than the linear range for aluminum oxide chromatoplates. However, the greater linear range for SiO_2 was not substantial. The Kubelka-Munk theory, and thus the Goldman theory, assume uniform particle size of the scattering medium. The commercial chromatoplates employed in Hurtubise's work probably had a range of particle sizes and more work is needed to determine the magnitude of the effect of particle size as related to the Goldman theory. Kortum (2) has discussed the dependence of scattering coefficient on particle size and states it is nearly inversely proportional to the average particle size. The aluminum oxide chromatoplates employed in Hurtubise's work had sX values greater than the silica gel chromatoplates. For both silica gel and aluminum oxide chromatoplates the sX values were substantially greater than the kX values. The data suggested that the average particle size for the silica gel chromatoplates was larger than the average particle size for the aluminum oxide chromatoplates. Particle size appeared to be a major factor contributing to the smaller 7/30 sXkX and 4/30 sXkX products for silica gel chromatoplates.

The experimental data presented by Hurtubise appear to be the only published data to support an analytical theory on solid surface luminescence analysis. He emphasized the need for more work under a variety of experimental and instrumental conditions to establish the validity of Goldman's equations and other theories. The results of the work indicated that under certain conditions Goldman's simplified equations can be applied to solid surface luminescence analysis to predict approximately at what point a calibration curve will become nonlinear and the approximate range of linearity of a calibration curve. Also the results showed that it is more advantageous to measure reflected fluorescence because of the greater relative fluorescence response except in those cases where the fluorescence quantum efficiency and layer thickness are such that reflected fluorescence and transmitted fluorescence give approximately the same response.

POLLAK AND BOULTON'S EQUATIONS

Pollak and Boulton (12) and Pollak (18,20) have developed a theory
of fluorescence for thin-layer chromatograms based on the Kubelka-
Munk theory and an electrical transmission line model. In a later
series of papers Pollak (19-22) developed the theoretical instrumental aspects of solid surface fluorescence analysis. These concepts
will be considered later in this chapter.

Pollak and Boulton (12) used the assumptions of the Kubelka-
Munk theory as discussed earlier and took into consideration both
excitation and fluorescent radiation in scattering media, as did
Goldman (11). Pollak and Boulton considered the effect of fluorescence on the propagation of exciting radiation as tantamount to
an increase in absorbance, namely, an additional energy loss of the
exciting radiation (12). Also, they obtained a differential equation that was essentially identical to an equation describing an
electrical transmission line with purely resistive parameters.
They used the electrical transmission line as a model to simulate
the optical behavior of turbid media on both absorption and fluorescence modes (5,12). Pollak (23) has considered the transmission
line model in some detail.

Pollak (18,20) presented two complex equations based on the
Kubelka-Munk theory and the electrical transmission line model.
The equations described the fluorescence reaching the surface on
the same side as the exciting radiation (reflection mode), and the
fluorescence reaching the side opposite the exciting radiation (transmission mode). He did not give solutions for the equations, but
discussed simplified equations which lend themselves to interpre-
tation in practical laboratory situations. These simplified equations are discussed below.

Media with strong scattering and relatively large absorbance
are often encountered in solid surface fluorescence analysis. Pollak
(18,20) presented two theoretical equations that are applicable
under these conditions for reflected and transmitted fluorescence.
He assumed the coefficient of fluorescence was relatively small

in both cases. The coefficient of fluorescence was defined as the
ratio of the intensity of fluorescence excited at any particular
point in the interior of the medium divided by the total energy density
of the exciting radiation at that point. The two equations are
given below for transmitted fluorescence and reflected fluorescence,
respectively (18,20). The layer thickness X does not appear in equations (3) and (4) because it was normalized by setting X = 1.

$$I_{FT} = \frac{Fc}{2} \exp(-\gamma_F) \left(1 - \varrho_E^2\right) \left(1 + \varrho_F\right) E_0 \qquad (3)$$

$$I_{FR} = \frac{Fc}{2} \frac{\left(1 - \varrho_E^2\right) \left(1 + \varrho_F\right)}{\gamma_E + \gamma_F} E_0 \qquad (4)$$

where the indices E and F refer to excitation radiation and fluorescent radiation, respectively, I_{FT} and I_{FR} are the intensities of
the fluorescent radiation emitted in the transmission mode and reflection mode, respectively, c is the amount of fluorescent substance
in the illuminated area, E_0 is the intensity of the exciting beam,
F is the coefficient of fluorescence of the component of interest,
ϱ is the coefficient of reflectance of a sheet of medium thick enough
so that its transmission can be disregarded, and γ is the natural
logarithm of the transmittance of a very thin sheet of medium, which
has negligible reflectance. Pollak (20) pointed out that ϱ^2 is so
small that chance variation of its magnitude can be neglected. The
influence of the terms containing ϱ were combined into a constant
correction factor b. Thus, equations (3) and (4) reduce to (20):

$$\frac{I_{FT}}{E_0} = \frac{Fcb}{2} \exp{-\gamma_F} \qquad (5)$$

$$\frac{I_{FR}}{E_0} = \frac{Fcb}{2} \frac{1}{\gamma_E + \gamma_F} \qquad (6)$$

It can be seen that I_{FT}/E_0 is a function of γ_F, but I_{FT}/E_0 is a
function of γ_E and γ_F. Pollak (20) made the conclusion that the
linear range for reflected fluorescence is more limited than the
linear range for transmitted fluorescence due to the influence of
γ_E. Pollak (18) also commented that the reflected fluorescence
is expected to be greater than the transmitted fluorescence. Pollak
did not expand further on the above comments.

COMPARISON OF GOLDMAN'S EQUATIONS AND POLLAK AND
BOULTON'S EQUATIONS

It would be nice at this point to make a detailed comparison of
the two sets of theoretical equations discussed in the previous
three sections. However, Goldman and Pollak and Boulton gave no experimental data to support their theoretical conclusions, and a very
limited amount of theoretical data was presented by the authors.
It appears the only experimental data obtained to support a theory
on solid surface luminescence analysis was obtained by Hurtubise
(17). He discussed Goldman's equations. Zweidinger and Winefordner
(24) derived expressions using the Kubelka-Munk theory and gave
supporting theoretical and experimental data that related phosphorescence intensity of species in optically inhomogeneous matrices to
various parameters. They were interested mainly in snow matrices
or densely cracked glasses that were formed at liquid nitrogen temperature. However, the concepts and equations developed by them should
prove useful in solid surface luminescence analysis.

It appears that Goldman's equations are more readily related
to optical theories because they were derived directly from the
Kubelka-Munk theory. Pollak and Boulton also employed the Kubelka-
Munk theory, but in addition used the properties of an electrical
transmission line as a model to simulate the optical behavior of
radiation in scattering media. Without more work both theoretically
and experimentally a firm conclusion cannot be made as to which
approach is more applicable to solid surface fluorescence analysis.
There is also the possibility that both approaches will have to be
modified extensively.

Goldman (11) predicted that at low amounts of fluorescer the
range of linearity for reflected fluorescence would be approximately
twice that of transmission in terms of kX. Pollak (20) predicted
that the range of linearity for reflected fluorescence would be
more limited than that for transmitted fluorescence because of the
influence of γ_E. The fluorescence data of Hurtubise (17) for silica
gel and aluminum oxide chromatoplates showed that the range of linearity for fluoranthene for both reflected and transmitted fluorescence

were relatively close. It seems obvious that there is now a need
to obtain fluorescence data for several compounds using several chromatographic systems so adequate comparisons can be made between Goldman's
equations and Pollak and Boulton's equations.

Finally, no published work has appeared on theoretical equations that describe calibration curves for room temperature phosphorescence from solid surfaces. It seems the equations described in
the preceding sections and the equations derived by Zweidinger and
Winefordner (24) should be applicable in this area.

THEORETICAL CONSIDERATIONS IN SOLID SURFACE
LUMINESCENCE INSTRUMENTATION

A limited amount of work has been done on the theoretical instrumental aspects of solid surface luminescence analysis. Certainly
many of the instrumental concepts developed by others on solution
luminescence analysis are applicable to solid surface luminescence
analysis (25,26). Reference 25 is particularly good in this respect.
However, because highly scattering media are employed in luminescence from solid surfaces, there are additional instrumental aspects
to consider, and instruments designed for solution luminescence
are not necessarily the best for solid surface luminescence analysis.
In fact, present commercial instruments designed to measure luminescence from solid surfaces can be improved to measure luminescence
more efficiently.

Goldman and Goodall (4) developed a theory for absorption methods
in thin-layer chromatography emphasizing the nonlinear relationship
between the amount of substance for analysis and absorbance in the
transmission mode. Later they applied the flying spot technique
for measurements in the transmission mode (14,15). Pollak (19)
considered the background theory for the design of instruments that
could be employed in the absorption mode and fluorescence mode for
scattering media. The following concepts apply mainly to the absorption mode. The intensity of source radiation from a scattering
medium can be spatially nonuniform, and the photodetector acts as

an integrator which averages the signal intensity over the whole
area it "sees." Also the signal intensity of source radiation from
the scattering media generally varies in a nonlinear fashion with
the spatial concentration of the measured substance in the scattering
media (4,5). Because of these factors the photodetector output
signal is a unique function of the amount of substance for analysis
only if the amount of the substance is constant over the whole area
"seen" by the detector. This particular condition can be very closely
approximated experimentally by making the area "seen" by the photodetector very small with respect to the zone of material on a scattering media such as a thin-layer chromatoplate. To completely
monitor the whole area of interest the chromatoplate must move in
two dimensions with respect to the medium. This is achieved by
the flying spot technique. A good discussion of this general approach has been presented for dual-wavelength point zig-zag scanning
of zones on thin-layer chromatoplate as a tool for quantitative assay
by Yamamoto et al. (27).

As mentioned, the preceding discussion applies mainly to the
absorption mode, and Pollak (19) emphasized that arguments in favor
of flying spot scanning do not necessarily apply to fluorescence
measurements. This is so because the fluorescent radiation received
by the photodetector is a linear function of the amount of substance
for analysis at small amounts of the substance (12,18). Goldman
(11) reached similar theoretical conclusions for small amounts of
the substance. This is in contrast to absorption methods in which
the photodetector output signal is in general a nonlinear function
of the spatial concentration of the substance. Both transmitted
and reflected signals in the absorption mode require linearization
in terms of the amount of substance for analysis except at extremely
small amounts of the substance (19,28). In the fluorescence quenching approach, linearization is required similar to transmittance
measurements (19), and two-dimensional scanning is recommended (11).
Fluorescence quenching is considered inferior to direct fluorescence
measurement because of additional optical noise provided by the

fluorescent medium, reduced sensitivity, and the need for double-beam scanning (11,20).

With direct fluorescence measurements from a scattering medium, the fluorescing substance stands out as a bright zone on a dark background. Theoretically the dark background yields almost an ideally flat baseline. However, electrical noise and residual fluorescence from impurities in the scattering medium can cause a jittery baseline. Electrical noise can be minimized by integrating the output signal over a long period of time (19,20). Because the residual fluorescence from impurities is almost time invariant it would not be diminished by integration. Pollak (20) suggested two integration approaches for fluorescent signals of very low intensities. These are integration on photographic film and photon counting with subsequent integration. Neither approach has found extensive application in luminescence from solid surfaces. Of the two approaches, photon counting with subsequent integration is more appealing. This is because of the difficulty in processing photographic film and because the reciprocity law of photography breaks down at very low light levels. In general the reciprocity law states that the degree of blackening of photographic film is proportional to the product of exposure time and illuminating intensity (20,29). With photon counting techniques it should be possible to extend detection limits of luminescent components to much lower levels than can be achieved presently (20,25,30).

Even though direct fluorescence measurements can give a nearly ideally flat baseline optical noise can be a problem. Optical noise is caused by random fluctuations of the optical transfer in the scattering medium. It is usually the principal factor which limits performance in solid surface luminescence analysis. Electrical noise becomes important at very low light levels and is generated in the photodetector and preamplifier stages. Several sources of optical noise are listed below (19,20).

1. Any specular component of the scattered radiation at the surface of medium varies randomly and as a result the intensity of the light entering the medium exhibits random

fluctuations from point to point. These fluctuations appear
as optical noise in both the transmission and reflection
modes.

2. In the reflection mode, it is possible for part of the
specularly reflected luminescence to reach the photodetector
and cause optical noise. This can be reduced by careful
optical design.

3. One serious source of optical noise is caused by local variations in the thickness of the medium.

4. Density fluctuations of the medium and nonuniform particle
size can cause optical noise.

5. Special treatment of the medium such as with chemical reagents can cause nonrandom changes in optical parameters.

6. Optical noise can also originate in the source radiation
through lamp instability.

Pollak (20) stated that theoretically the signal-to-noise ratio
has a definite value which does not depend on the coefficient of
fluorescence (F) or the amount of fluorescent component [See equations (3) and (4)]. He estimated the signal-to-noise ratio for
transmitted fluorescence and reflected fluorescence, and the relation-
ships are given in the following:

$$\left(\frac{S}{N}\right)_{FT} \approx \frac{1}{\delta_\gamma} \tag{7}$$

$$\left(\frac{S}{N}\right)_{FR} \approx \frac{\gamma}{\delta_\gamma} \tag{8}$$

The optical noise fluctuations are contained in the coefficients
γ which were defined with equations (3) and (4). The term δ_γ represents the rms value of the fluctuations of the γ terms. Pollak
assumed that γ_E and γ_F were approximately equal [equation (6)].
Equations (7) and (8) indicate that signal-to-noise ratio is both
constant and independent of the amount of the fluorescer. However,
the output signal of the photodetector is still dependent on the
stability of the excitation source.

According to equations (3) and (4), increasing the intensity
of the excitation radiation will increase the fluorescence intensity

of a fluorescent component. This does not necessarily mean the sensitivity will increase, because there may be a corresponding increase in the fluorescence of impurities in the scattering media. However, the increased output of fluorescent radiation can mask electrical noise and improve sensitivity and accuracy (20).

Pollak (20) has discussed double-beam scanning for fluorescence from solid surfaces. The goal of double-beam scanning in fluorometry is different from that in absorption work. In absorption work, the main goal is to obtain a smooth baseline. Noise produced by incremental changes in the amount of absorber in the sample beam (incremental zone noise) is not reduced. With fluorescence work, baseline noise can be very low and the main concern is reduction of incremental zone noise. Pollak (20) gives the signal-to-noise ratio for scanning the double-beam transmission mode by the following:

$$\frac{S}{N} \simeq \frac{1 + Q_F}{\delta Q_F} \tag{9}$$

The term Q_F is the coefficient of reflection, and δQ_F is the rms value of the fluctuations of the coefficients Q_F. Pollak assumed the wavelength of the reference beam is close to that of the fluorescence wavelength of the fluorescer. According to equation (9) the improvement obtainable depends on the magnitude of the term Q_F. A typical value for Q_F is 0.35, and for δQ_F a typical value is 0.035. These values give S/N of approximately 39. For single-beam fluorometry a typical S/N value is about 10. Thus it appears theoretically a double-beam system is better than a single-beam system.

Pollak proposed some useful and interesting theoretical concepts for the instrumental aspects of solid surface fluorescence analysis. However, Pollak presented no experimental data in support of the conclusions. As with the theoretical equations of Goldman and Pollak that describe the relationship between fluorescence intensity and amount of fluorescer, there is a need for experimental data to substantiate theoretical instrumental concepts. Some of the data can be obtained with available commercial instrumentation; however, there is a need for new instrumentation to help in substantiating or

modifying the theoretical concepts. Also, because room temperature
phosphorescence is now analytically useful, theoretical and practical
consideration should be given to the instrumental aspects of this
important area. Certainly the work of Winefordner and others has
initiated this but more work on the instrumental aspects of room
temperature phosphorescence is needed (31-34). Two areas that should
be investigated that are of particular importance and applicable
to both fluorescence and phosphorescence are excitation of the sample
and detection of the emitted radiation. Turnable lasers should prove
beneficial for excitation of the sample, and detector systems that
have high collection efficiency for luminescence are particularly
important with highly scattering media.

EFFECT OF NONUNIFORM CONCENTRATION DISTRIBUTION WITH DEPTH

A rigorous mathematical derivation of the change of luminescence
with variation of the concentration profile with the depth in the
scattering medium is very difficult. Pollak (21) has derived equations for two extreme situations in fluorescence analysis. For the
first case, it was assumed the material of interest was concentrated
at the far side of exciting radiation. For the second case, it
was assumed the material was concentrated at the near side of exciting radiation. It is important to recall that it is assumed
normally that the material of interest is uniformly distributed
throughout the medium. Pollak compared the equations for the two
extreme cases with equations for uniform density profile. He came
to the theoretical conclusion that fluorescence determinations in
the transmission mode yield results that are almost independent
from the distribution of the analyzed material with depth. Also,
he concluded that fluorescence measurements in the reflection mode
are strongly dependent on distribution of concentration. Pollak's
conclusions suggest that if there is a variable distribution of
concentration in the depth of the scattering medium fluorescence
transmittance measurements are preferred.

Pollak gave no experimental data to support the conclusions, and it is important to consider the magnitude of the effect of non-uniform concentration distribution with depth. It may be that the magnitude is unimportant in many situations such as with strong fluorescers. As with other theoretical conclusions, in solid surface luminescence analysis very little experimental data have been collected in support of these conclusions.

THE RELATIONSHIP BETWEEN THE DIMENSIONS OF THE SCATTERING MEDIUM AND SENSITIVITY

Pollak (22) considered theoretically the decrease of the zone area of a fluorogen and its effect on sensitivity. Sensitivity was defined as the smallest change of the optical signal which can be reliably identified as not being of purely random origin. Under the condition where the signal-to-noise ratio is constant and independent of concentration, namely, sufficiently above the sensitivity threshold, reduction of zone area does not offer any major advantages. In fact, for best results Pollak states the zone area should be large and subject to the restriction that optical noise prevails over electrical noise. The main reason for this paradoxical conclusion is that with a constant signal-to-noise ratio the smoothing effect of integration increases when zone area is made larger. Also, Pollak commented, the improvement of the sensitivity threshold is proportional to the square root of the decrease in zone area. Pollak gave no experimental data to support the above conclusions.

Pollak (22) discussed the effect of layer thickness on sensitivity. The following equations relate fluorescence intensity in the transmission (A_{TF}) and reflection (A_{RF}) modes, respectively (22).

$$A_{TF} \simeq \frac{cF}{2} \exp(-X\gamma) \tag{10}$$

$$A_{RF} \simeq \frac{cF}{2} \frac{1}{2X\gamma} \tag{11}$$

The term X = layer thickness, and the other terms were defined earlier in this chapter. From equations (10) and (11), it is concluded that

in both cases the amplitude of a fluorescent signal increases if
X becomes smaller. This assumes that cF remains constant and the
scattering medium does not absorb exciting radiation. Both modes
are susceptible to optical noise because of thickness variations.
It seems that reducing X is a way of increasing the sensitivity and
improving the accuracy near sensitivity threshold concentrations.
Pollak presented no experimental data to support his conclusions.
Goldman's equations [see equations (1) and (2)] indicate that fluorescence intensity is proportional to X as long as the expressions
7/30 sXkX and 4/30 sXkX are small, and the scattering medium does
absorb exciting radiation. Also, Goldman's transmission equation
[equation (1)] does not have an exponential term as does Pollak's
transmission equation [equation (10)]. With Goldman's equations
if one assumes a constant amount of fluorescer in the linear range,
but the layer thickness decreases, then the fluorescence intensity
will increase in both the reflected and transmitted modes. This
is true because the absorption coefficient k increases under these
conditions. Thus, one can arrive at a similar conclusion with Pollak's
and Goldman's equations, namely, as layer thickness decreases, fluorescence intensity increases with the same amount of fluorescer.

Goldman (11) has given equations for reflected and transmitted
fluorescence in which the scattering medium itself absorbs exciting
radiation. In both cases, the fluorescence intensity varies inversely
as the layer thickness increases.

Goldman and Pollak give no experimental evidence for their conclusions. As emphasized many times, it is important to substantiate
these important theoretical conclusions. Only with a blending of
the theoretical and experimental will the best conditions for solid
surface luminescence be achieved.

REFERENCES

1. P. Kubelka, *J. Opt. Soc. Amer.*, *38*, 448 (1948).

2. G. Kortum, *Reflectance Spectroscopy*, Springer-Verlag, New York,
    1969.

3. W. W. Wendlandt and H. G. Hecht, *Reflectance Spectroscopy*, Interscience Publishers, New York, 1966.

4. J. Goldman and R. R. Goodall, *J. Chromatogr.*, *32*, 24 (1968).

5. V. Pollak and A. A. Boulton, *J. Chromatogr.*, *50*, 19 (1970).

6. V. Pollak and A. A. Boulton, *J. Chromatogr.*, *50*, 39 (1970).

7. V. Pollak, *J. Chromatogr.*, *105*, 279 (1975).

8. H. G. Hecht, *Anal. Chem.*, *48*, 1775 (1976).

9. F. A. Huf, H. J. DeJong, and J. B. Schute, *Anal. Chim. Acta*, *85*, 341 (1976).

10. F. A. Huf, *Anal. Chim. Acta*, *90*, 143 (1977).

11. J. Goldman, *J. Chromatogr.*, *78*, 7 (1973).

12. V. Pollak and A. A. Boulton, *J. Chromatogr.*, *72*, 231 (1972).

13. G. Kortum, *Reflectance Spectroscopy*, Springer-Verlag, New York, 1969.

14. J. Goldman and R. R. Goodall, *J. Chromatogr.*, *40*, 345 (1969).

15. J. Goldman and R. R. Goodall, *J. Chromatogr.*, *47*, 386 (1970).

16. R. R. Goodall, *J. Chromatogr.*, *78*, 153 (1973).

17. R. J. Hurtubise, *Anal. Chem.*, *49*, 2160 (1977).

18. V. Pollak, *Optica Acta*, *21*, 51 (1974).

19. V. Pollak, *J. Chromatogr.*, *123*, 11 (1976).

20. V. Pollak, *J. Chromatogr.*, *133*, 49 (1977).

21. V. Pollak, *J. Chromatogr.*, *133*, 195 (1977).

22. V. Pollak, *J. Chromatogr.*, *133*, 199 (1977).

23. V. Pollak, *I.E.E.E. Trans. Bio-Med. Eng.*, *17*, 287 (1970).

24. R. Zweidinger and J. D. Winefordner, *Anal. Chem.*, *42*, 639 (1970).

25. J. D. Winefordner, S. G. Schulman, and T. C. O'Haver, *Luminescence Spectrometry in Analytical Chemistry*, Wiley-Interscience, New York, 1972.

26. D. M. Hercules, ed., *Fluorescence and Phosphorescence Analysis*, John Wiley & Sons, Inc., New York, 1966.

27. H. Yamamoto, T. Kurita, J. Suzuki, R. Hira, K. Nakano, H. Makage, and K. Shibata, *J. Chromatogr.*, *116*, 29 (1976).

28. V. Pollak, *Opt. Acta*, *23*, 25 (1976).

29. E. D. Olsen, *Modern Optical Methods of Analysis*, McGraw-Hill Book Company, New York, 1975, p. 268.

30. A. J. Diefenderfer, *Principles of Electronic Instrumentation*, W. B. Saunders Co., Philadelphia, Penn., 1972.

31. T. Vo-Dinh, G. L. Walden, and J. D. Winefordner, *Anal. Chem.*, *49*, 1126 (1977).

32. E. L. Yen-Bower and J. D. Winefordner, *Appl. Spectrosc.*, *33*, 9 (1979).

33. E. L. Yen, G. D. Boutilier, and J. D. Winefordner, *Can. J. Spectrosc.*, *22*, 120 (1977).

34. C. D. Ford and R. J. Hurtubise, *Anal. Chem.*, *51*, 659 (1979).

5

Some Interactions Responsible for
Room Temperature Phosphorescence

As discussed in Chapter 3, room temperature phosphorescence (RTP)
is an expanding new area of research. It is an important area in
chemical analysis because solution room temperature phosphorescence
is not useful analytically and low-temperature phosphorescence techniques can be cumbersome (1,2). It is possible today to adsorb an
organic compound onto a solid surface and obtain analytically useful
RTP data (3,4). It has been shown that room temperature phosphorescence is both a sensitive and a selective analytical tool (3,5).
Because of the newness of room temperature phosphorescence, many
theoretical and practical aspects remain to be developed. In this
chapter some of the evidence and proposed mechanisms of interactions
that result in RTP of organic compounds will be considered.

INTERACTIONS ON FILTER PAPER

Schulman and Parker (6) considered the effects of moisture, oxygen,
and the nature of the support-phosphor interaction employing two
model compounds on several supports, but mainly filter paper was
considered as a solid support. Earlier, Schulman and Walling (7,8)
suggested that surface adsorption of phosphorescent compounds to
the support inhibits collisional deactivation of the triplet state
and restricts oxygen quenching when the sample is dried rigorously.

Schulman and Parker (6) extended the above hypothesis by proposing that hydrogen bonding of ionic organic molecules to hydroxyl groups on the solid surface is the primary mechanism of providing the rigid sample matrix for RTP. They also proposed that moisture acts to disrupt hydrogen bonding and aids in the transport of O_2 into the sample matrix. Their work showed that both moisture and oxygen can independently quench RTP. This was indicated by relative intensity data for humidified argon and oxygen presented in Table 5.1 and Fig. 5.1 for samples adsorbed on Whatman No. 1 filter paper. In Fig. 5.1 NaBPCA refers to sodium 4-biphenylcarboxylate. In the absence of oxygen, moisture acts alone as a powerful quenching agent (Fig. 5.1). At low humidities a moderate quenching effect was noted, and at high humidities quenching was quite dramatic

TABLE 5.1 Relative Intensities of Sodium 4-Biphenylcarboxylate Samples in Ar and O_2 as a Function of Relative Humidity

% H^a	I_{Ar}^b	Relative standard deviation in Ar^c	$I_{O_2}^d$	Relative standard deviation in O_2^c	$Q_{H_2O}^e$	$Q_{O_2}^f$
0.0	100	1.2	70.9	1.5	0.0	29.1
3.2	98.1	1.1	70.3	1.3	1.9	27.8
8.5	91.6	2.4	61.3	1.9	8.4	30.3
18.0	57.6	1.2	25.5	3.8	42.4	32.1
37.1	12.8	2.6	2.5	14.0	87.2	10.3
58.3	2.6	7.1	0.4	2.9	97.4	2.2
80.5	0.7	4.3	0.1	17.0	99.3	0.6
100.0	0.3	11.0	0.0		99.7	0.3

[a] Percent relative humidity of gas at 298 K.

[b] Intensity in argon relative to 0% humidity in argon (I_{Ar}^o).

[c] Calculated from triplicate runs and given in percent.

[d] Intensity in oxygen relative to I_{Ar}^o.

[e] $I_{Ar}^o - I_{Ar}$.

[f] $I_{Ar} - I_{O_2}$.

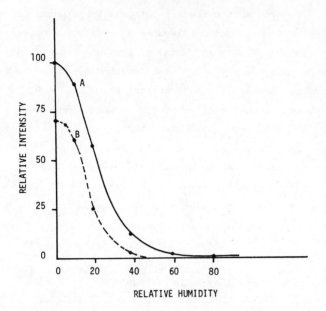

FIG. 5.1 Relative intensities of NaBPCA samples in Ar (curve A = I_{Ar}, ——) and O_2 (curve B = I_{O_2}, - - - -) as a function of humidity.
(Reprinted with permission from E. M. Schulman and R. T. Parker, *J. Phys. Chem., 81*, 1932 (1977). Copyright by the American Chemical Society.)

(Fig. 5.1). On the basis of previous work (7,8) and more recent work (6), Schulman and Parker concluded that moisture competes with surface hydroxyl functions for hydrogen bonding to the phosphor molecules and ties up hydroxyl groups so the phosphor is not held rigidly. In other words, water "softens" the matrix, allowing collisional deactivation. Quenching by triplet ground-state oxygen occurred in the absence of moisture, but the magnitude of oxygen quenching was facilitated greatly by the presence of moisture. This is indicated in Fig. 5.2. The term Q_{O_2} represents the amount of quenching due only to oxygen at a given humidity, and the term I_{Ar} represents the phosphorescence intensity in argon relative to 0% humidity in argon. Schulman and Parker (6) concluded that moisture must be regarded as the most important contributor to quenching RTP because it can both transport oxygen into the sample matrix

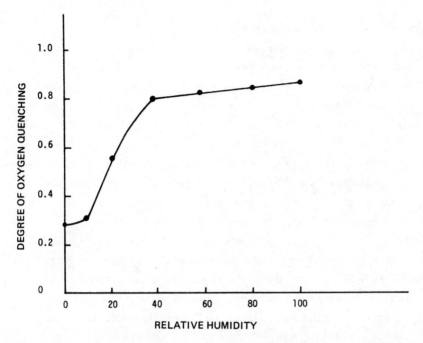

FIG. 5.2 Degree of O_2 quenching, represented by Q_{O_2}/I_{Ar} plotted
as a function of relative humidity for NaBPCA samples. (Reprinted
with permission from E. M. Schulman and R. T. Parker, *J. Phys. Chem.*,
81, 1932 (1977). Copyright by the American Chemical Society.)

and allow normal collisional deactivation to operate. Schulman
and Parker (6) employed alkaline solutions of two model compounds,
sodium 4-biphenylcarboxylate and sodium 1-naphthoate. Schulman
and Walling (7,8) found that organic acid phosphors give a much weaker
RTP than their corresponding sodium salts, although both are capable
of hydrogen bonding to the hydroxylic surface. These results and
the results of other researchers indicate that polar or ionic organic compounds have the greater probability of giving RTP (3,4,
6-9,15,16). However, there is a need to investigate the importance
of ionic interactions of organic compounds in RTP more extensively. De
Lima and de M. Nicola (10) gave data on the effect of NaOH concentration
on RTP for several compounds, but no complete investigation on this
aspect has yet appeared. Wellons, Paynter, and Winefordner (11)

compared the phosphorescence signals of several compounds adsorbed
on filter paper at room temperature and at 77 K in a solid matrix.
The signal ratios are given in Table 5.2 The data in Table
5.2 are indicative of the degree of rigidity in which the molecules
were held on filter paper versus a solid solution at 77 K. The
authors commented that the molecules that have the most ionic sites
show the greatest rigidity. Vanillin and 2,4-dithio-pyridimidine are
doubly charged species in strong alkaline solution. Uncharged species
like sulfaquanidine and 5-acetyl-uracil give the signals at room
temperature. These two uncharged compounds gave relatively strong
phosphorescence signals. The results from the previous two compounds
indicated hydrogen bonding was important in holding the molecules
rigid. For the other compounds ionic interactions are important.
Vo-Dinh, Yen, and Winefordner (12) reported on the RTP of several
polyaromatic hydrocarbons on paper support. Heavy-atom solvents
such as silver nitrate and sodium iodide were found to induce phosphorescence emission from these compounds. The previous example

TABLE 5.2 Comparison of Phosphorescence Signals at Room Temperature
and at 77 K

Compound[a]	Max. phos. signal at 77 K / Max. phos. signal at 330 K
4-Amino-benzoic acid	6
6-Methylmercapto-purine	8
2-Amino-6-methylmercapto-purine	19
4-Amino-2,6-dihydroxyl pyrimidine	50
2,4-Dithio-pyridimidine	1.8
Sulfaquanidine	110
Tryptophane	10
5-Acetyl-uracil	220
2-Thio-6-amino-uracil	26
Vanillin	1.6

[a] All compounds prepared in 1 M NaOH. Concentrations are approxi-
5 mM.

Reprinted with permission from Ref. 11.

shows that nonionic compounds can give useful RTP analytical information.

Niday and Seybold (13) investigated matrix effects on the lifetime of RTP. They measured the phosphorescence half-life of 2-naph-
thalenesulfonate on filter paper with various compounds added to
the filter paper. Some of the compounds added were NaF, NH_4Cl,
CH_3COONa, H_3BO_3, glycine, glucose, and sucrose. In every case examined, an increase in the phosphorescent lifetime was observed.
In no case was the phosphorescent lifetime as long as that observed
at 77 K in a rigid mixture of ethyl ether, isopentane, and ethyl
alcohol in a ratio of 5:5:2 (EPA). The authors commented that the
common explanation for RTP is that the matrix holds the adsorbed
compound rigid and thereby restricts vibrational motions necessary
for nonradiative decay from the triplet state. They stated their
results were consistent with the above model by assuming that packing
the matrix with salts and sugars further inhibits internal motion
of the phosphorescent compound. The added compounds "plug up" the
channels and interstices of the matrix, decreasing oxygen permea-
bility, and protecting the phosphorescent molecules from quenching
by oxygen.

INTERACTIONS ON SODIUM ACETATE

Von Wandruszka and Hurtubise (4) investigated the interactions responsible for RTP of compounds adsorbed on sodium acetate. Comparisons of molecular structures and consideration of reflectance, fluorescence, and infrared spectra in addition to surface-area data and
solvent considerations allowed the postulation of certain molecular
criteria for RTP. The interactions of p-aminobenzoic acid (PABA)
with sodium acetate were considered in detail. The interactions
involved the formation of the sodium salt of PABA as well as hydrogen
bonding. Their data indicated that the molecule was adsorbed flatly
on sodium acetate.

Table 5.3 compares phosphorescence intensities of some of the
compounds investigated by von Wandruszka and Hurtubise (4). They

TABLE 5.3 Phosphorescence Intensities of Compounds Adsorbed on
Sodium Acetate[a]

Compound	77 K	Room Temp (330 K)	330 K/77 K
PABA	1000	333	0.333
p-Hydroxybenzoic acid	1040	48.0	0.046
3-Methyl-4-aminobenzoic acid	232	29.0	0.125
N,N-Dimethyl-4-aminobenzoic acid	260	42.0	0.16
Benzocaine	146	--	--
Terephthalic acid	238	0.9	0.0038
Hydroquinone	38	6.8	0.18
Folic acid	132	43.0	0.32
p-Hydroxymandelic acid	56	6.5	0.116
p-Aminohippuric acid	1200	262	0.218
5-Hydroxyindoleacetic acid	200	32.0	0.160
5-Hydroxytryptophan	210	39.0	0.186

[a] 77 K phosphorescence of PABA arbitrarily set at 1000 units. All
compounds 500 ng/10 mg NaOAc.

Reprinted with permission from R. M. A. von Wandruszka and R. J.
Hurtubise, *Anal. Chem.*, *49*, 2164 (1977). Copyright by the American
Chemical Society.

used ethanol solutions of the compounds to deposit the compounds
on the sodium acetate. The data imply that differences in RTP intensities are not due to inherent molecular effects, but rather
to differences in rigidity of the adsorbed species. PABA appeared
to be adsorbed most strongly, because it retained relatively the
greatest fraction of RTP. They found that certain molecular require-
ments were needed to obtain strong RTP signals. For example, 3-methyl-
4-aminobenzoic acid led to a 11.5-fold reduction in RTP compared
to PABA. From their results they concluded that the presence of
a carboxyl attached to the 1 position appeared to be one requirement
for compounds with the benzene nucleus. Also, attached to the 4
position on the ring, an electron-donating, hydrogen-bonding substi-
tuent appeared to be necessary. Two compounds with the indole nucleus showed RTP on sodium acetate, 5-hydroxyindoleacetic acid

(5-HIAA), and 5-hydroxytryptophan (5-HTP). Useful signals were
obtained only when alkaline ethanolic solutions of these compounds
were evaporated onto sodium acetate. Von Wandruszka and Hurtubise
found that the solvent employed in adsorbing the compounds on sodium
acetate was important in obtaining RTP. Ethanol and n-propanol
could be employed, but water gave a three-fold reduction in RTP
for PABA. Aprotic solvents such as ether, acetone, dimethylformamide,
and cyclohexane gave no RTP of PABA on sodium acetate.

Von Wandruszka and Hurtubise (4) postulated that the adsorption
of PABA on sodium acetate was preceded by partial neutralization
with dissolved sodium acetate in alcoholic solutions. The PABA
anion formed had a strong tendency to adsorb on the surface, forming
the sodium salt. This was supported by the strong RTP of the sodium
salt of PABA when it adsorbed on sodium acetate suspended in an
ethanolic solution of the sodium salt of PABA. No RTP signals were
observed for PABA and the sodium salt of PABA dissolved in acetone
or dimethylformamide. They commented that probably ion pairing
and the formation of conjugated species in these solvents were responsible for the lack of RTP. They also hypothesized that the
formation and adsorption of anions also occurs with other compounds
that phosphoresce on sodium acetate. In 5-HIAA and 5-HTP solutions,
it appeared that the dissolved acetate did not give the required
neutralization of the solute because 5-HIAA and 5-HTP are very weak
acids. Thus it was necessary to add the stronger base NaOH to ethanol
solutions of these compounds. Data were also presented that showed
about 84% of PABA was converted to the sodium salt on the sodium
acetate surface.

Von Wandruszka and Hurtubise (4) studied the mode of adsorption
of PABA and other compounds on sodium acetate, talc, and starch
with the aid of reflectance spectroscopy. The compounds investigated
did not give RTP on talc and starch. Sodium acetate is a polar surface and interacted strongly with most of the compounds investigated.
The reflectance spectra of PABA adsorbed on sodium acetate, starch,
and talc are shown in Fig. 5.3. As indicated, there was a hypsochromic

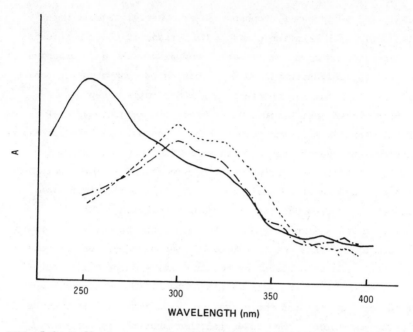

FIG. 5.3 Reflectance spectra of PABA adsorbed on sodium acetate
(———), starch (————), and talc (-·-·). (Reprinted with permission
from R. M. A. von Wandruszka and R. J. Hurtubise *Anal. Chem., 49,*
2164 (1977). Copyright by the American Chemical Society.)

shift of 35 nm in the reflectance maximum of PABA on sodium acetate
as compared to PABA on talc or starch. This result indicated strong
interactions between sodium acetate and adsorbed PABA, suggesting
the formation of the sodium salt of PABA upon adsorption. Other
supporting reflectance and fluorescence data were also presented
by these authors (4).

   The mode of adsorption of PABA on sodium acetate was further
investigated with infrared spectroscopy by von Wandruszka and Hurtubise
(4). The infrared spectra of PABA, of sodium acetate and of PABA
adsorbed on sodium acetate were obtained. Strong sodium acetate
infrared bands obscured much of the adsorbate spectra, but a number
of observations and conclusions were made. The N-H stretching vibra-
tions of PABA at 3350-3450 cm^{-1} disappeared for PABA adsorbed on
sodium acetate. This indicated that the bands were shifted to longer

wavelengths and broadened due to hydrogen bonding between the amino group and the carboxyl group of the surface sodium acetate. The infrared spectrum of a similarly prepared o-aminobenzoic acid sample adsorbed on sodium acetate retained its N-H infrared bands. There is an intramolecular hydrogen bond between the amino group and the carboxyl group in o-aminobenzoic acid and thus this compound apparently is not held to the surface by strong intermolecular hydrogen bonding. o-Aminobenzoic acid did not give RTP.

Von Wandruszka and Hurtubise (4) assumed that chemisorbed PABA molecules on the sodium acetate surface are distinguished from physically adsorbed molecules by their RTP behavior. Only those molecules that strongly and directly interact with the sodium acetate surface are held rigidly enough to show RTP. They postulated that PABA molecules in the second and subsequent adsorbed layers would not phosphoresce, but instead would decrease the signal by adsorbing exciting radiation and possibly emitted phosphorescence from the chemisorbed layer. It was found that the maximum RTP signal was obtained at 6100 ng PABA on 10 mg sodium acetate. They considered the maximum RTP signal to correspond to complete monolayer coverage. Von Wandruszka and Hurtubise (4) used a method developed by Snyder (14) to calculate the surface area occupied by a flatly adsorbed PABA molecule. They assumed PABA was adsorbed flatly based on the spectral and other data they obtained. They then calculated the surface area of sodium acetate to be 1.8 m^2/g. The previous value was identical to the surface area of sodium acetate obtained commercially. Their results suggest strongly that PABA is adsorbed flatly on sodium acetate. Further calculations showed that two sodium acetate molecules were needed to hold one PABA molecule. A similar study was done with 5-HIAA, and it was found that three sodium acetate molecules were needed to hold one 5-HIAA molecule. These ideas are illustrated schematically in Fig. 5.4. The results obtained by von Wandruszka and Hurtubise may be used as a guide for investigation of other molecules and surfaces in order to give RTP a still broader analytical scope.

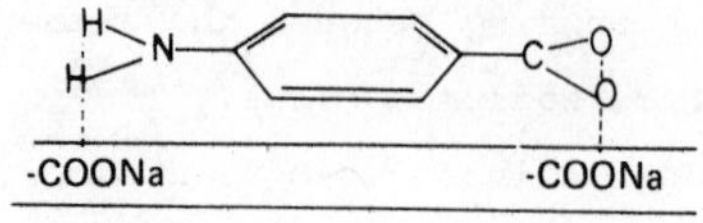

-COONa   -COONa

(a)

-COONa   -COONa   -COONa

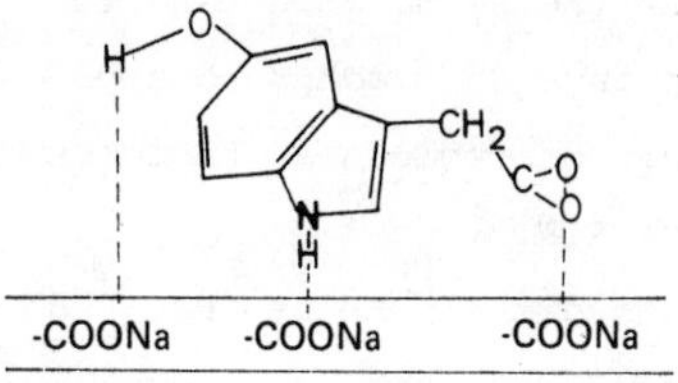

(b)

FIG. 5.4   Adsorption of PABA (a) and 5-HIAA (b) on a sodium acetate
surface.   (Reprinted with permission from R. M. A. von Wandruszka
and R. J. Hurtubise, *Anal. Chem.*, *49*, 2164 (1977). Copyright by
the American Chemical Society.)

INTERACTIONS ON SILICA GEL

Ford and Hurtubise (15) studied the interactions of the phthalic
acid isomers and other compounds on silica gel. They obtained RTP
from several compounds by spotting ethanol solutions of the compounds
on chromatoplates and then drying the chromatoplates. Neither acidic
nor alkaline solutions were employed in their work. As with several
other workers, they found it was necessary to dry the sample to obtain strong RTP signals.

For RTP to occur it is necessary that the adsorbed molecules
be held rigidly to the surface. Compounds with structural similari-
ties were studied by Ford and Hurtubise to gain some insight into
the mechanism of surface interaction between the adsorbate and silica
gel. The structures of some of the compounds studied are given
below.

Terephthaldehyde and p-diacetylbenzene exhibited no RTP on
silica gel after drying. Both of these compounds exhibited moderate
phosphorescence on silica gel at liquid nitrogen temperature. Terephthalamide gave a weak RTP signal when adsorbed on dried silica
gel and a relatively weak signal at liquid nitrogen temperature.

TPA — Isophthalic acid — Phthalic acid — Terephthaldehyde — p-diacetylbenzene — Terephthalamide

Terephthalic acid (TPA) gave strong RTP signals when adsorbed on
silica gel. Because terephthaldehyde and p-diacetylbenzene did
not exhibit RTP, and terephthalamide and TPA did, the interaction
of the latter two compounds with silica gel is stronger compared
to the interaction of the former two compounds with silica gel.
Terephthalamide contains two primary amide nitrogens as well as amide
hydrogens that are not found in terephthaldehyde or p-diacetylbenzene.
TPA contains two hydroxyl groups which are absent in the dialdehyde
and diacetyl compounds. The presence of primary amide nitrogens
and amide hydrogens or hydroxyl groups allows for stronger interaction with the silica gel surface. Ford and Hurtubise (15) postulated
that the main type of interaction was hydrogen bonding between the
surface hydroxyl groups of the silica gel surface and the primary
amide nitrogens and amide hydrogens or hydroxyl groups on the molecules.

    Ford and Hurtubise (15) excited TPA adsorbed on dried silica
gel with shortwave ultraviolet radiation after the silica gel chromatoplate was submerged in n-hexane. They assumed n-hexane would
not disrupt hydrogen bonding between the adsorbate and adsorbent.
Their assumption was valid because RTP from the submerged chromatoplate was observed with little or no loss of intensity. When the
same dried chromatoplate was placed in distilled water, the RTP
of TPA was immediately quenched, presumably because of the disruption

of hydrogen bonds between TPA and silica gel surface. The RTP re-
appeared after the chromatoplate was removed from the water and dried.
They examined the other phthalic acid isomers to determine what
role para-substitution played on the surface interaction. Phthalic
acid on dried silica gel exhibited moderate RTP, and isophthalic
acid exhibited a very weak RTP signal under similar conditions.
The data suggested that the strongest interaction with the silica
gel surface is provided by para-substitution, and meta-substitution
provided the weakest interaction. They also noted that no RTP signals were obtained by drying the chromatoplate first and then put-
ting the sample on the dried chromatoplate. This suggested that
water and ethanol are adsorbed very rapidly and occupy sites on
silica gel that would otherwise be available to other molecules.

Ford and Hurtubise (15) compared the low-temperature phosphorescence signals with the RTP signals for several compounds. They
assumed that the compound with the strongest interaction with silica
gel would exhibit the least amount of signal enhancement at low
temperature. PABA had a signal enhancement of 5.3 at low temperature
compared to its RTP signal, and terephthalamide had an enhancement
of 11.0 compared to its RTP signal. These results indicated that
the interaction of terephthalamide with silica gel was weaker than
the interaction of PABA with silica gel. This simple experimental
approach is a rapid way of obtaining a general measure of the strength
of interaction of molecules with a solid surface at room temperature.

The RTP of pyridine type nitrogen heterocycles adsorbed on
silica gel was investigated by Ford and Hurtubise (16). Benzo[f]-
quinoline was used as a model compound, and it was studied on several
silica gel samples. Luminescence, reflectance, and infrared spectroscopy were used to obtain a better understanding of the analytical
conditions needed for strong RTP. The results showed that silica
gel chromatoplates containing a polymeric binder with carboxyl groups
were the best samples for inducing strong RTP from benzo[f]quinoline.
It was shown that the polymer itself was essential for inducing
strong RTP. Infrared spectroscopic results and other information

strongly suggested a salt polyacrylic acid was used as a binder
at about 5% by weight. Their results clearly indicated that enhanced
RTP signals were obtained only when the binder was in its acidic
form. Thus, acidic solutions of the nitrogen heterocycles had to
be spotted on the chromatoplates for strong RTP signals. Ford and
Hurtubise proposed hydrogen bonding between the pi-electron systems
of the protonated nitrogen heterocycles and the carboxyl groups of
the polymeric binder. Since the acidic polymer is a stronger acid
than the silanol groups from silica gel, the carboxyl groups would
be expected to form stronger hydrogen bonds with the pi-electron
system of nitrogen heterocycles and therefore hold the compounds
more rigidly to the surface. In addition the protonated nitrogens
of the nitrogen heterocycles could form hydrogen bonds with the
carbonyl oxygens of the carboxyl groups of the binder. Ford and
Hurtubise also showed that polyacrylic acid-NaCl mixtures induced
strong RTP from nitrogen heterocycles indicating that silica gel
was not necessary for inducing RTP. As additional evidence for
the hydrogen bonding mechanism, the hydrocarbon analogue of benzo[f]-
quinoline, phenanthrene, was adsorbed on a 5% polyacrylic acid-NaCl
mixture from ethanol solution. After drying the mixture, the phenanthrene exhibited a strong RTP. Phenanthrene also exhibited a moderate
RTP signal when spotted from an acidic ethanol solution onto silica
gel chromatoplates which contained the salt of an acidic binder.
Their results suggested the excess acid converted carboxylate groups
to carboxyl groups which then interacted strongly with phenanthrene.
Because phenanthrene has no heteroatom, they concluded the main
interaction responsible for the enhanced RTP observed from this compound was hydrogen bonding between the carboxyl groups of the binder
and the pi-electron system of phenanthrene.

Ford and Hurtubise (15) also used silica gel chromatoplates
that contained the salt of an acidic binder in their work with phthalic
acid isomers which was discussed earlier in this section. It seems
that in addition to the hydrogen bonding mechanism proposed by them
that the salt of the acidic binder could interact with the phathalic

acid isomers by a simple acid-base reaction to hold the isomers
rigidly to the surface. However, no data has been presented to support this.

HEAVY-ATOM EFFECT

Researchers have shown that the heavy-atom effect can be useful
analytically (12,17-24). Very little research has been done on
the theoretical aspects of the heavy-atom effect related to RTP.
In the presence of heavy atoms, the fluorescence intensity may be
reduced substantially or even quenched completely, and the phosphorescence may be substantially increased. Heavy atoms chemically bonded
to the luminescent molecule enhance quenching by increasing spin-
orbital coupling, which favors intersystem crossing. This is called
the *internal heavy-atom effect*. It has been proposed that heavy
atoms in a solvent affect spin-orbital coupling in a solute because
the solute is surrounded by a tightly bound solvent sheath (24).
This is called the *external heavy-atom effect*.

White and Seybold (24) studied the effect of added halide salts
on the room temperature fluorescence and phosphorescence of 2-naph-
thalene sulfonate adsorbed on dried filter paper. They observed a
normal external heavy-atom effect on luminescence, and found the
phosphorescence intensity to increase strongly in the order NaF <
NaCl < NaBr < NaI. They found that the external heavy atoms increase
the radiative decay constant k_p more than the competing nonradiative
constant k_{qp}.

Bower and Winefordner (21) studied the RTP of several compounds
with emphasis on polynuclear aromatic hydrocarbons (PAH). They found
the external heavy-atom effect led to significant enhancement of
RTP for PAH. The trend for the cations studied was $Tl^+ > Ag^+ > Pb^{2+} >$
$Hg^{2+} > Tl^+$, which resulted in enhanced spectral features of the emission bands. The authors concluded that more studies involving greater
variation in conditions were needed before an exact mechanism could
be elucidated. However, Bower and Winefordner speculated that silver(I)

was bonded to the PAH forming a π complex. Also, silver(I) could
interact with functional groups on the paper. They suggested the
hydroxyl groups from cellulose participated in the bonding, and
thus a link between the analyte molecule and support was developed,
providing the necessary rigidity for observation of phosphorescence.
They also believed silver(I) not only provided heavy-atom effect
but provided a rigid environment for the phosphorescing molecule.
They assumed that restriction of translational, vibrational, and
rotational movements, and resistance to penetration by oxygen were
responsible for the lack of susceptibility to quenching. They postulated that moisture would serve to weaken the matrix rigidity and
allow oxygen to be dissolved in the solvent and thus quench the
phosphorescence. Schulman and Parker (6) have made a similar proposi-
tion concerning oxygen quenching.

REFERENCES

1. J. D. Winefordner, S. G. Schulman, and T. C. O'Haver, *Lumines-
    cence Spectrometry in Analytical Chemistry*, Wiley-Interscience,
    New York, 1972.

2. D. M. Hercules, ed., *Fluorescence and Phosphorescence Analysis*,
    John Wiley & Sons, Inc., New York, 1966.

3. R. A. Paynter, S. L. Wellons, and J. D. Winefordner, *Anal.
    Chem., 46*, 736 (1974).

4. R. M. A. von Wandruszka and R. J. Hurtubise, *Anal. Chem., 49*,
    2164 (1977).

5. R. M. A. von Wandruszka and R. J. Hurtubise, *Anal. Chem., 48*,
    1784 (1976).

6. E. M. Schulman and R. T. Parker, *J. Phys. Chem., 81*, 1932 (1977).

7. E. M. Schulman and C. Walling, *Science, 178*, 53 (1972).

8. E. M. Schulman and C. Walling, *J. Phys. Chem., 77*, 902 (1973).

9. M. Roth, *J. Chromatogr., 30*, 276 (1967).

10. C. G. de Lima and E. M. de M. Nicola, *Anal. Chem., 50*, 1658
     (1978).

11. S. L. Wellons, R. A. Paynter, and J. D. Winefordner, *Spectrochim.
     Acta*, Part A, *30*, 2133 (1974).

12. T. Vo-Dinh, E. L. Yen, and J. D. Winefordner, *Talanta, 24*, 146
     (1977).

13. G. L. Niday and P. G. Seybold, *Anal. Chem., 50*, 1577 (1978).

14. L. R. Snyder, *Principles of Adsorption Chromatography*, Marcel Dekker, Inc., New York, 1968, p. 199.

15. C. D. Ford and R. J. Hurtubise, *Anal. Chem., 50*, 610 (1978).

16. C. D. Ford and R. J. Hurtubise, *Anal. Chem., 52*, 656 (1980).

17. P. G. Seybold and W. White, *Anal. Chem., 47*, 1199 (1975).

18. J. L. McHale and P. G. Seybold, *J. Chem. Educ., 53*, 654 (1976).

19. T. Vo-Dinh, E. L. Yen, and J. D. Winefordner, *Anal. Chem., 48*, 1186 (1976).

20. I. M. Jakovljevic, *Anal. Chem., 49*, 2048 (1977).

21. E. L. Yen-Bower and J. D. Winefordner, *Anal. Chim. Acta, 102*, 1 (1978).

22. E. L. Yen-Bower and J. D. Winefordner, *Anal. Chim. Acta, 101*, 319 (1978).

23. J. J. Aaron and J. D. Winefordner, *Analysis, 7*, 168 (1979).

24. W. White and P. G. Seybold, *J. Phys. Chem., 81*, 2035 (1977).

6

Some Analytical
Procedural Considerations

In this chapter the application of the sample to solid surfaces
for luminescence measurements will be emphasized. Other procedural
details, especially for thin-layer chromatography (TLC) with subsequent quantitation of separated components, have been treated elsewhere (1-4). The techniques developed by Guilbault and co-workers
for the assay of enzymes, substrates, activators, and inhibitors
were discussed in Chapter 3. In the chapters on applications that
follow, consideration will be given to additional procedural details.

For luminescence measurements from solid surfaces in thin-layer
chromatography the size of the initial spot should be kept as uniform
as possible. If a syringe is used a source of error during application of the sample is "creep back" on the tip of the syringe (3).
Part of the drop can curl back around the tip of the syringe and
remain after the sample is placed on the surface. This source of
error can be minimized by using a very fine tip and by coating the
tip with silicone (3). Generally, as will be seen in the following
chapters, these special precautions are not necessary in most applications. Samuels and Fisher (5) used folded nichrome wire loops
to apply nanoliter volumes to flat solid surfaces with coefficients
of variations of 1-2% for the cumulative application of 1 µl. De
Silva, Berkersky, and Puglisi (6) taking no special precautions,
used commercial silica gel chromatoplates, a calibrated 50-µl

disposable pipette to spot 9-acridanones, measured reflected fluorescence of the 9-acridanones, and obtained the following mean and
standard deviation values for peak area of three 9-acridanones, respectively: 0.71 ± 0.06 cm^2, 0.87 ± 0.07 cm^2, 2.34 ± 0.16 cm^2. The
mean values are each from three individual measurements and the
amount applied to the plate was 1 ng. MacMullan and Heveran (7)
stated that using standard techniques to apply samples to TLC chromatoplates and assuming a random distribution of error, the precision
obtained was ± 3-5% for the application of 10 µl. In a recent paper,
Issaq and Barr (8) commented that micropipettes, microsyringes,
or capillary tubes are especially useful in quantitative TLC work,
and the devices dispense from 1 to 25 µl of sample with a precision
of ± 1% or less. Touchstone and Dobbins (2) have discussed in some
length the application of the sample to TLC chromatoplates. Also,
they have considered manual sample applicators, automatic sample
applicators, and aids and guides for manual sample application.
Touchstone, Levin, and Murawec (9) emphasized that highly concentrated areas of the applied spot should be avoided in quantitative
TLC. Under these conditions, the solvent tends to flow around rather
than through a spot, resulting in uneven sample diffusion. Also,
spot travel of at least an R_F of 0.25 and not greater than 0.75
should be accomplished in quantitative TLC work because the area
per unit of solute is more uniform than those of higher or lower
R_F values (2).

     In room temperature phosphorescence (RTP) work, the application
of the sample to the solid surface is also important. The ideas
discussed previously for TLC are applicable generally to RTP. A
popular solid surface employed in RTP is filter paper. In most work
with RTP the sample is applied to the solid surface, and then phosphorescence is measured. Thus far chromatographic separations have
not been carried out with subsequent measurements of RTP. Vo-Dinh,
Walden, and Winefordner (10) discussed the shape of the sample spot
on filter paper. With their system, a filter paper roll with cir-
cular sample spots on it moved at a constant speed, and phosphorescence emission was recorded with a modified Aminco-Bowman

spectrophotofluorometer. Samples were delivered drop by drop by
manually spotting 3 μl of solution onto the moving filter paper with
a hypodermic syringe. PABA and chrysene were employed as model
compounds. In both cases, the shapes of the peaks were symmetric,
but PABA showed two maxima at the border of the curve. This indicated that a major part of the PABA diffused toward the outer border
of the spot. Chrysene stayed mostly in the middle part of the spot.
PABA was dissolved originally in a solution of NaOH:NaI, and chrysene
in an ethanol/water (1/1; v/v) solution containing 0.1 M of $AgNO_3$.
The authors emphasized that for quantitative analysis care must
be taken because it is necessary to integrate the area under each
curve to obtain an accurate measure of the amount of various compounds.

Later, Yen-Bower and Winefordner (11) discussed the reproducibility for RTP with an improved continuous sampling system. Their
continuous sampling system was discussed in Chapter 3. They obtained
increased reproducibility (5% relative standard deviation) compared
to their previous systems which gave approximately 15% relative
standard deviation (11). The increased reproducibility was due to
the ease of positioning the support in the sample holder and the
fewer handling steps needed compared to the previous systems. The
researchers obtained two distinct peaks for carbazole, and for this
phenomenon they again recommended integration of the area under
the peaks in quantitative studies.

Von Wandruszka and Hurtubise (12) applied various samples to
sodium acetate homogeneously for RTP studies. A 25-μl volume of
ethanol was introduced into a 4 × 0.4 cm test tube from a micropi-
pette, and 1- to 6-μl volumes of standard or sample solution were
added from a 10-μl Hamilton syringe. A fixed amount of sodium acetate was added to each tube with a measuring spoon that had the
same volume as the depressions in a special sample plate they designed. The tube was placed in an oven at 80°C until the ethanol
was evaporated. The dry solid was transferred quantitatively to
a small mortar and pestle which was used to gently break up conglom-
erate particles. The powder was then transferred to the depressions

in the special sample plate and smoothed over with a spatula. For
a series of 10 PABA samples at 100 ng each, the relative standard
deviation for the RTP measurements was 5%. For a similar series
at 10 ng each the relative standard deviation was 7%.

Ford and Hurtubise (13) applied acidic ethanol solutions of
nitrogen heterocycles to silica gel or filter paper for RTP studies.
All compounds were dissolved initially in distilled ethanol prior
to spotting on a silica gel chromatoplate or filter paper. Each
solution was acidified by mixing 40 µl of concentrated HCl with
5 ml (final volume) of the ethanol to be spotted. The acid was added
by means of a Quick-pipette (Helene Labs, Beaumont, Texas). Prior
to spotting, the plates or filter papers were marked with a pencil
so straight dashed lines appeared in the same plane across the solid
surface. The Hamilton syringe containing the adsorbate solution
was positioned carefully between dashes so the center of each spot
lay in line with the markings. Best reproducibility was realized
when spot size was maintained as constant as possible. Of the 3
µl spotted, one-third of the solution was placed initially on the
plate or filter paper. Care was taken not to increase the size
of this initial spot while spotting the remaining two-thirds of
the sample solution. With some practice this technique gave good
reproducibility. After spotting the silica gel chromatoplate, the
aluminum backed plate was allowed to dry in the atmosphere for 5
min and was then placed in an oven at 105-110°C to dry for an additional 30 min. The plate was removed from the oven and allowed
to cool to room temperature. Then the plate was positioned carefully
on the stage of a densitometer. The RTP signal was maximized for
the most concentrated spot, and the signals for all the spots were
measured by starting the moving stage of the densitometer. It was
not necessary to maximize the signal for each spot because of the
careful spotting technique employed.

When filter paper was employed as the adsorbent, 3 µl of the
solution containing the nitrogen heterocycle was placed on the paper,
again using a Hamilton syringe. The filter paper was allowed to

dry in the atmosphere for 5 min before being placed in an oven to
dry for 10 min at 105-110°C. If the paper was placed in the oven
immediately after spotting, the spots of the compound which were
acidified appeared blackened after drying. After drying, the paper
was placed on the stage of the spectrodensitometer and the RTP signals were recorded as described above for silica gel plates (13).

For reproducibility data, Ford and Hurtubise (13) spotted 16
spots of benzo[f]quinoline (B[f]Q) from a 0.1 M HCl ethanol solution
at 100 ng each onto a silica gel chromatoplate. The RTP was maximized for the most concentrated spot, and the signals obtained as
described above. The standard deviation was determined to be 1.50
relative intensity (R.I.) units. The 95% confidence limits were
calculated to be 46.5 ± 0.8 R.I. units. A similar experiment with
4-azafluorene resulted in a standard deviation of 0.93 R.I. units
and 95% confidence limits of 16.7 ± 0.5 R.I. units for sixteen 100
ng spots. Ford and Hurtubise also investigated plate-to-plate variation of RTP. Three aluminum-backed silica gel chromatoplates were
each spotted with five 100 ng spots of B[f]Q. After drying, the
RTP signals were recorded for each spot with a spectrodensitometer.
The average RTP intensities obtained for the three plates were 62.0
(plate 1), 54.8 (plate 2), and 57.6 (plate 3). Because of the plate-
to-plate variation in RTP of compounds adsorbed on silica gel, they
recommended that any quantitative method which employs RTP from
compounds adsorbed on commercial chromatoplates should include RTP
measurement of the proper standards along with the unknowns on the
same chromatoplate. They did not run similar experiments with filter
paper.

Bethke, Santi, and Frei (14) discussed the use of a data pair
technique that allows the chromatographer to compensate for varying
layer thickness or edge effects in TLC. This technique is an internal compensation approach and works by pairing up the measurements
of two spots on the same chromatoplate--one spot from the edge and
the other spot from the center. A chromatoplate with n spots yields
n/2 data pairs. They found that the improvement in reproducibility

was highly significant. This approach was applied to ultraviolet
reflectance spectrometry and could be applied to solid surface luminescence analysis. Later Bethke and Frei (15) developed a transferable calibration technique and applied the calibration transfer
model to ultraviolet reflectance measurements and in situ fluorescence measurements from the chromatoplates. A fundamental problem
with quantitative fluorescence measurements in TLC is the need to
run standards on the same chromatoplate as the unknown sample because of changes in plate characteristics and in chromatographic
conditions. Bethke and Frei (15) addressed this problem with their
transferable calibration technique. The approach involved first
obtaining the average slope ($\bar{b}$) of calibration curves or a relative
average slope ($\overline{b_{rel}}$) from several different chromatoplates. Then
for a new chromatoplate the calibration curve was obtained by one
calibration value and computing the intercept for the calibration
curve. For example, the peak area A_x (calibration value) was determined for a known amount of component C_x. From these two values
and with either the average slope value ($\bar{b}$) or the relative average
slope ($\overline{b_{rel}}$), the intercept a_i was calculated for the new calibration curve. Table 6.1 compares the two calibration methods by the
mean relative error of 36 analyses of digoxin ampule solutions.

TABLE 6.1 Mean Values of the Relative Errors of 36 Analyses of
Digoxin Ampules (ΔU_{rel})[a] (Value After Transferred Calibration)

Transferred value	Three separated groups (%)	Total series (%)
$\bar{b}$	1.8	2.9
$\overline{b_{rel}}$	0.7	0.7

[a] $(\Delta U_{rel}) = (U_{transf.\ calibr.} - U_{indiv.\ calibr.})/U_{indiv.\ calibr.} \times 100$.

The digoxin solutions were analyzed by direct fluorescence evaluation on chromatoplates after derivatization of digoxin. The results for the relative analytical error ΔU_{rel} indicated no significant difference for the serial transfer or the total transfer (Table 6.1). Also the results showed that the relative analytical error is smaller if $\overline{b}_{rel}$ is employed in the transferable calibration technique. Bethke and Frei's approach should be useful when a large number of samples have to be analyzed necessitating the use of several chromatoplates.

Rulon and Cardone (16) described photographic techniques for fluorescent spots on thin-layer chromatoplates. Frei (17) reported a critical study of some parameters in quantitative in situ thin-layer densitometry. Touchstone, Schwartz, and Levin (18) determined phenobarbital and parahydroxylated phenobarbital and stated their work could serve as a general guide for those wishing a basic understanding of the principles of thin-layer densitometry.

REFERENCES

1. J. C. Touchstone, ed., *Quantitative Thin Layer Chromatography*, John Wiley & Sons, Inc., New York, 1973.

2. J. C. Touchstone and M. F. Dobbins, *Practice of Thin Layer Chromatography*, John Wiley & Sons, Inc., New York, 1978.

3. J. G. Kirchner, *J. Chromatogr.*, *82*, 101 (1973).

4. J. C. Touchstone and J. Sherma, eds., *Densitometry in Thin Layer Chromatography*, John Wiley & Sons, Inc., New York, 1979.

5. S. Samuels and C. Fisher, *J. Chromatogr.*, *71*, 297 (1972).

6. J. A. F. de Silva, I. Berkersky, and C. V. Puglisi, *J. Pharm. Sci.*, *63*, 1837 (1974).

7. E. A. MacMullan and J. E. Heveran, in *Quantitative Thin Layer Chromatography*, J. C. Touchstone, ed., John Wiley & Sons, Inc., New York, 1973, pp. 213-233.

8. H. J. Issaq and E. W. Barr, *Anal. Chem.*, *49*, 83A (1977).

9. J. C. Touchstone, S. S. Levin, and T. Murawec, *Anal. Chem.*, *43*, 858 (1971).

10. T. Vo-Dinh, G. L. Walden, and J. D. Winefordner, *Anal. Chem.*, *49*, 1126 (1977).

11. E. L. Yen-Bower and J. D. Winefordner, *Appl. Spectrosc.*, *33*, 9 (1979).

12. R. M. A. von Wandruszka and R. J. Hurtubise, *Anal. Chem.*, *48*, 1784 (1976).

13. C. D. Ford and R. J. Hurtubise, *Anal. Chem.*, *51*, 659 (1979).

14. H. Bethke, W. Santi, and R. W. Frei, *J. Chromatogr. Sci.*, *12*, 392 (1974).

15. H. Bethke and R. W. Frei, *Anal. Chem.*, *48*, 50 (1976).

16. P. W. Rulon and M. J. Cardone, *Anal. Chem.*, *49*, 1640 (1977).

17. R. W. Frei, *J. Chromatogr.*, *64*, 285 (1972).

18. J. C. Touchstone, M. F. Schwartz, and S. S. Levin, *J. Chromatogr.*, *15*, 528 (1977).

7

Room Temperature
Phosphorescence Applications

Because of the newness of room temperature phosphorescence (RTP),
very little has been published on the application of RTP to a variety
of "real-life" samples. The emphasis has been on obtaining important
analytical data such as reproducibility, limits of detection, and
range of linearity of calibration curves. RTP has added a new dimension to solid surface luminescence analysis in that the potential
exists for measuring both fluorescence and phosphorescence of compounds adsorbed on solid surfaces at room temperature. Also, because low-temperature conditions are not needed, phosphorescence
measurements are obtained more easily. In addition, RTP is both
sensitive and selective. In this chapter a survey of the compounds
that have given analytically useful RTP signals and the solid supports used to obtain RTP signals will be considered. In addition,
the RTP determination of p-aminobenzoic acid (PABA) in vitamin tablets
will be discussed.

   Some solid surfaces that have been employed for RTP are cellulose (1), filter paper (1-3), silica gel (3-5), aluminum oxide
(3), sodium acetate (6-9), sucrose (10), starch (10), and polyethylene
fiber-paper blend (10). The most popular solid surface has been
filter paper, and Paynter, Wellons, and Winefordner (11) and Vo-Dinh,
Walden, and Winefordner (2) have evaluated several paper types.
Silica gel and sodium acetate have shown great promise as solid

surfaces (4-8). As work progresses in RTP, additional solid surfaces should become available. In the previous chapter details in
applying the sample to the surface were given. Vo-Dinh and
Winefordner (12) have reviewed some aspects of RTP as a method of
analysis.

Many drugs, pharmaceuticals, and compounds of biological interest have shown RTP. Generally the limits of detection for these
compounds range from subnanogram to submicrogram levels. Some RTP
spectral characteristics and analytical conditions for several compounds are given in Table 7.1. Several of the compounds listed gave

TABLE 7.1 Room Temperature Phosphorescence Characteristics and
Analytical Conditions for Drugs, Pharmaceuticals, and Compounds
of Biological Interest

Compound	Excitation wavelength (nm)	Emission wavelength (nm)	Experimental conditions	Ref.
4-Aminobenzoic acid	273	426	1 M NaOH in water, filter paper	13
2,4-Dithiopyrimidine	374	464		13
4-Amino-2,6-dihydroxyl-pyrimidine	310	421		13
2-Amino-6-methylmercapto-purine	331	487		13
2,6-Diaminopurine	286	451		13
6-Methylpurine	268	449		13
6-Chloropurine	280	463		13
6-Methylmercaptopurine	292	466		13
Sulfanilamide	267	426		13
Sulfquanidine	267	426		13
Tryptophan	280	448		13
5-Acetyluracil	304	421		13
2-Thio-6-aminouracil	307	441		13
4-Hydroxyl-3-methyloxy-benzaldehyde (vanillin)	332	519		13

TABLE 7.1 (Continued)

Compound	Excitation wavelength (nm)	Emission wavelength (nm)	Experimental conditions	Ref.
Adenine	290	470	1 M NaOH - 1 M NaI in water, filter paper	14
Guanine	280	450		14
6-Chloropurine	290	460		14
6-Methylmercaptopurine	290	458		14
L-(-)-Tryptophan	290	448		14
Sulfanilamide	280	427		14
Cocaine hydrochloride	285	460		14
PABA	280	430		14
Salicylic acid	320	470		14
Barbituric acid	290	455		14
Antrenyl	280	500	Aqueous solution[a]	15
	280	500	+1 M NaOH[a]	
	280	450, 490	+1 M NaOH + 1 M NaI	
	280	500	+0.2 M $AgNO_3$[a], filter paper	
Anturane	305	500	Ethanolic solution[a]	15
	335	515	+1 M NaOH[a]	
	285	460	+1 M NaOH + 1 M NaI	
	315	515	+0.2 M $AgNO_3$[a], filter paper	

TABLE 7.1 (Continued)

Compound	Excitation wavelength (nm)	Emission wavelength (nm)	Experimental conditions	Ref.
Apresoline	290	450	Aqueous solution[b]	15
	280	440	+1 M NaOH	
	280	440	+1 M NaOH + 1 M NaI	
	290	500	+0.2 M $AgNO_3$[a], filter paper	
Butazolidin	300	500	Ethanolic solution[a]	15
	300	500	+1 M NaOH[a]	
	300	500	+1 M NaOH + 1 M NaI[a]	
	300	500	+0.2 M $AgNO_3$[a], filter paper	
Epinerphrine	280	500	130 ppm in 1 M NaOH[a]	15
	280	500	+1 M NaI[a], filter paper	
Esidrex	290	490	Ethanolic solution[a]	15
	275	440	+1 M NaOH	
	275	440	+1 M NaOH + 1 M NaI	
	275	500	+0.2 M $AgNO_3$[a], filter paper	

TABLE 7.1 (Continued)

Compound	Excitation wavelength (nm)	Emission wavelength (nm)	Experimental conditions	Ref.
Forhistal	290	500	Aqueous solution[a]	15
	290	500	+1 M NaOH[a]	
	310	500	+1 M NaOH + 1 M NaI[a]	
	310	500	+0.2 M AgNO$_3$[a], filter paper	
Locorten	305	500	Ethanolic solution[a]	15
	275	500	+1 M NaOH[a]	
	275	455, 490	+1 M NaOH + 1 M NaI[a]	
	275	515	+0.2 M AgNO$_3$[a], filter paper	
Metamphetamine hydrochloride	250, 270	500	In 10% ethanol[a]	15
	250, 270	500	+1 M NaOH[a]	
	250, 270	500	+1 M NaOH + 1 M NaI[a]	
	250, 270	500	+0.2 M AgNO$_3$, filter paper	
Metandren	280	490	Ethanolic solution[a]	15
	300	490	+1 M NaOH[a]	
	300	490	+1 M NaOH + 1 M NaI[a]	
	300	510	+0.2 M AgNO$_3$[a], filter paper	

TABLE 7.1   (Continued)

Compound	Excitation wavelength (nm)	Emission wavelength (nm)	Experimental conditions	Ref.
Metopirone	300	500	Ethanolic solution[a]	15
	338	510	+1 M NaOH[a]	
	285	460	+1 M NaOH + 1 M NaI	
	300	500	+0.2 M AgNO$_3$[a], filter paper	
Nupercaine hydrochloride	295	445	Aqueous solution	15
	295	470	+1 M NaOH[a]	
	295	445	+1 M NaI	
	260	450	+1 M NaOH + 1 M NaI[a]	
	295	515	+0.2 M AgNO, filter paper	
Otrivine	275	490	Aqueous solution[a]	15
	275	510	+1 M NaOH[a]	
	335	515	+1 M NaOH + 1 M NaI[a]	
	310	515	+0.2 M AgNO$_3$[a], filter paper	
Priscoline hydrochloride	280	500	Aqueous solution[a]	15
	280	440, 490	+1 M NaOH[a]	
	280	500	+1 M NaOH + 1 M NaI	
	280	490	+1 M NaI[a]	
	280	500	+0.2 M AgNO$_3$[a], filter paper	

TABLE 7.1   (Continued)

Compound	Excitation wavelength (nm)	Emission wavelength (nm)	Experimental conditions	Ref.
Privine hydrochloride	280	490, 510	Aqueous solution	15
	280	435, 490, 520	+1 M NaOH	
	280	435, 490, 520	+1 M NaOH + 1 M NaI	
	280	435, 490, 520	+1 M NaI	
	280	515	+0.2 M $AgNO_3$, filter paper	
Procaine hydrochloride	305	445	Aqueous solution	15
	275	430	+1 M NaOH	
	275	430	+1 M NaOH + 1 M NaI	
	300	460	+0.2 M $AgNO_3$, filter paper	
Regitine hydrochloride	310	500	Aqueous solution[a]	15
	310	500	+1 M NaOH[a]	
	280	490, 515	+1 M NaOH + 1 M NaI[a]	
	280	500	+0.2 M $AgNO_3$[a], filter paper	
Ritalin hydrochloride	280	500	Aqueous solution[a]	15
	332	515	+1 M NaOH	
	282	495, 525	+1 M NaOH + 1 M NaI	
	282	500	+1 M NaI[a]	
	300	510	+0.2 M $AgNO_3$[a], filter paper	

TABLE 7.1   (Continued)

Compound	Excitation wavelength (nm)	Emission wavelength (nm)	Experimental conditions	Ref.
Serpasil (reserpine)	302	500	Ethanolic solution[a]	15
	315	510	+1 M NaOH[a]	
	280	495, 520	+1 M NaOH + 1 M NaI	
	280	500	+1 M NaI[a]	
	302	500	+0.2 M $AgNO_3$[a], filter paper	
Sitrom	300	500	Ethanolic solution[a]	15
	300	500	+1 M NaOH[a]	
	300	500	+1 M NaOH + 1 M NaI[a]	
	300	500	+0.2 M $AgNO_3$[a], filter paper	
Tandearil	300	500	Ethanolic solution[a]	
	300	500	+1 M NaOH[a]	
	300	500	+1 M NaOH + 1 M NaI[a]	
	300	510	+0.2 M $AgNO_3$[a], filter paper	
Tegretol	310	500	Ethanolic solution[a]	15
	310	500	+1 M NaOH[a]	
	285	495, 520	+1 M NaOH + 1 M NaI[a]	
	310	515	+0.2 M $AgNO_3$[a], filter paper	

TABLE 7.1   (Continued)

Compound	Excitation wavelength (nm)	Emission wavelength (nm)	Experimental conditions	Ref.
Thiopropazate dihydrochloride	300	515	Aqueous solution[a]	15
	300	520	+1 M NaOH	
	275	490, 515	+1 NaOH + 1 M NaI	
	280	490, 515	+1 M NaI[a]	
	280	510	+0.2 M AgNO$_3$, filter paper	
Tofranil	300	500	Aqueous solution[a]	15
	335	510	+1 M NaOH[a]	
	280	450	+1 M NaOH + 1 M NaI	
	310	500	+0.2 M AgNO$_3$[a], filter paper	
Trasentine hydrochloride	280	500	Aqueous solution[a]	15
	335	510	+1 M NaOH[a]	
	275	490, 520	+1 M NaOH + 1 M NaI	
	300	490, 520	+1 M NaI	
	300	515	+0.2 M AgNO$_3$[a], filter paper	
Vinblastine sulfate	305	500	Aqueous solution[a]	15
	278	490, 515	+1 M NaOH	
	300	490	+1 M NaOH + 1 M NaI[a]	
	278	490, 515	+1 M NaI	
	300	510	+0.2 M AgNO$_3$[a], filter paper	

TABLE 7.1 (Continued)

Compound	Excitation wavelength (nm)	Emission wavelength (nm)	Experimental conditions	Ref.
Vioform	300	500	Ethanolic solution[a]	15
	300	500	+1 M NaOH[a]	
	300	510	+1 M NaOH + 1 M NaI	
	300	510	+0.2 M AgNO$_3$[a]	
	288	500	+0.2 M AgNO$_3$ + 1 M HNO$_3$[a] filter paper	
p-Aminobenzoic acid (PABA)	290	426	Ethanol, sodium acetate	7
p-Aminobenzoic acid, sodium salt	290	426		7
p-Aminohippuric acid	328	448		7
p-Hydroxybenzoic acid	285	420		7
Folic acid	320	465		7
5-Hydroxyindoleacetic acid	312	510		7
5-Hydroxytryptophan	355	500		7
Benzamide of PABA	300	430		7
3-Methyl-4-aminobenzoic acid	290	430		7
N,N-Dimethyl-4-aminobenzoic acid	290	430		7
m-Aminobenzoic acid	290	430		7
Hydroquinone	285	430	[c]	7
p-Hydroxymandelic acid	290	420	[c]	7
Cinoxacin, 1-ethyl-1,4-dihydro-4-oxo-[1,3]di-oxolo[4,5-g]cinnoline-3-carboxylic acid	290, 370	515	Methanol, filter paper impregnated with lead tetraacetate	16

TABLE 7.1 (Continued)

Compound	Excitation wavelength (nm)	Emission wavelength (nm)	Experimental conditions	Ref.
Diphenyl	275	470	Methanol, filter paper impregnated with lead tetraacetate	16
Diphenylacetic acid	320	500		16
4-Diphenylcarboxylic acid	300	490		16
4,4'-Diphenyldisulfonic acid	290	485		16
1-Naphthalene sulfonic acid	290	490, 515		16
2-Naphthoic acid	290, 300	490, 515		16
PABA	300	430		16
Quinine sulfate	335	510		16
2-Fluorobiphenyl	280	460		16
1-Ethyl-1,4-dihydro-7-methyl-4-oxo-1,8-naphthyridine-3-carboxylic acid	265, 336, 338	472	NaOH or NaOH:NaI in water, filter paper	17
1-Ethyl-1,4-dihydro-4-oxo-1,8-naphthyridine-3,7-dicarboxylic acid	270, 357	470		17
7-(Acetylamino)-1-ethyl-1,4-dihydro-4-oxo-1,8-naphthyridine-3-carboxylic acid	282, 346	456		17
1-Ethyl-1,4-dihydro-7-hydroxy-4-oxo-1,8-naphthyridine-3-carboxylic acid	280, 320	468		17

TABLE 7.1 (Continued)

Compound	Excitation wavelength (nm)	Emission wavelength (nm)	Experimental conditions	Ref.
7-Methyl-1,8-naphthyridin-4-ol	250, 265, 353	474	NaOH or NaOH:NaI in water, filter paper	17

[a] The signal was not useful analytically and was not distinguishable from background luminescence.

[b] The signal was not useful analytically but was distinguishable from background luminescence.

[c] Very weak signals obtained.

analytically useful signals. However, as indicated by the footnotes in Table 7.1, under certain conditions some of the compounds gave weak RTP signals, and the signals were not useful analytically. Yen-Bower and Winefordner (15) commented that silver(I) ions had little or no effect as heavy-atom perturbers on the compounds they investigated; however, sodium iodide led usually to enhanced RTP and improved limits of detection. Also, compounds investigated by Yen-Bower and Winefordner with similar structures such as auturane and butazolidin did not respond similarly. They suggested that thermal decomposition with an infrared lamp during the drying step might be the cause of lack of RTP for some of the compounds. Several of these compounds were dried in a glove bag through which room temperature nitrogen was passed, but RTP was still not produced. Their results showed, however, that very low limits of detection could be obtained, mainly when a heavy-atom perturber such as sodium iodide was present.

De Lima and de M. Nicola (17) studied the effect of different parameters such as NaOH concentration, effect of irradiation time, and the effect of temperature in the sample compartment on the last five compounds listed in Table 7.1. They found that for each analyte there seemed to exist, at room temperature, an optimum range of

NaOH concentration where the NaOH-paper support provided the strongest
RTP signals. Interestingly, no effect on the phosphorescence signal
was observed at low temperature for all the compounds examined between 1×10^{-2} and 12 M NaOH. The compounds generally showed a
decrease in RTP with irradiation time. For these experiments, they
eliminated the infrared component of the excitation source by use
of a continuous flow, cool-water filter between excitation source
and the quartz window of the sample compartment. The authors concluded that a photochemical modification of the absorbed molecules
had occurred. Also, the sample compartment was heated gradually
($\sim$30-60°C) to study the effect of temperature on wavelength maxima
and on signal intensities. All the compounds showed a decrease in
RTP with an increase in temperature. With 1-ethyl-1,4-dihydro-7-
methyl-4-oxo-1,8-naphthridine-3-carboxylic acid, the decrease in
RTP was irreversible. This was probably due to an irreversible
thermal reaction. With the other compounds, the signal intensities
returned to the original values when the temperature was lowered
to room temperature.

Seybold and White (18) discussed the analytical advantages of
the external heavy-atom effect. For example, the use of external
heavy-atom effect can eliminate the need for an auxiliary phosphorescence assembly and increase phosphorescence intensity. Seybold and
White stated that, in effect, a perturbing heavy atom performs the
function of a mechanical chopper by distinguishing between fluorescence and phosphorescence signals. Also, as demonstrated by others,
the detection limits in RTP analysis can be improved substantially
with the external heavy-atom effect.

Vo-Dinh, Yen, and Winefordner (19) and Yen-Bower and Winefordner
(9) obtained RTP signals from polycyclic aromatic hydrocarbons (PAH).
The main significance of these results is that analytically useful
RTP signals have been obtained from nonionic compounds. Vo-Dinh,
Yen, and Winefordner (19) used filter paper as a solid surface,
1:1 v/v mixture of ethanol and water as a solvent, and heavy-atom
perturbers (NaI and $AgNO_3$) to induce phosphorescence emission from

PAH. With NaI only weak RTP signals were obtained from pyrene. Silver nitrate, however, proved to be very effective in enhancing the phosphorescence signals of pyrene. The RTP of carbazole was enhanced by 0.5 M NaOH:NaI (1:1). This was attributed partly to the ionic character of carbazole in a strongly basic solvent in which deprotonation at the nitrogen position could occur. Under these conditions, the ionic species could be rigidly adsorbed onto the substrate.

Later Bower and Winefordner (9) found that RTP could be obtained from several PAH by using Tl^+, Ag^+, Pb^{2+}, and Hg^{2+} ions to induce RTP. A comparison was made between sodium acetate and filter paper as solid surfaces for the compounds. Also, the effect of different gaseous environments was investigated. Table 7.2 compares the results obtained with sodium acetate and filter paper as solid surfaces and Ag^+ ion as a heavy atom. As indicated in Table 7.2, better limits of detection were obtained with filter paper. Table 7.3 compares the effect of different heavy atoms on the RTP of PAH. Thallium(I) was the greatest enhancer of RTP with the trend being $Tl^+ > Ag^+ > Pb^{2+} > Hg^{2+}$. Yen-Bower and Winefordner (9) found that flow gas used during the measurement step was important. For example, pyrene showed almost a threefold increase over air when nitrogen was used. The authors commented that the rate of signal growth reflected the rate of drying. This would depend on the porosity and thickness of the filter paper, the way it was mounted in the sample holder, the solvent employed, and the particular compound investigated.

Aaron, Kaleel, and Winefordner (20) made a comparative study of low-temperature and room temperature phosphorescence characteristics of 32 pesticides. Low-temperature phosphorimetry was shown to be sensitive with limits of detection ranging between 0.001 and 30 µg/ml. The RTP approach was simple and specific for some of the pesticides with absolute limits of detection between 10 and 50 ng. The researchers commented that the RTP technique is more suitable than conventional low-temperature phosphorimetry if one wants to

TABLE 7.2 Limits of Detection for Several Polyaromatic Hydrocarbons on Sodium Acetate Pellets and Filter Paper

Compound	Sodium acetate pellets			Filter paper		
	λ_{ex} (nm)	λ_{em} (nm)[a]	LOD[b] (ng)	λ_{em} (nm)	λ_{em} (nm)[a]	LOD[b] (ng)
Anthracene	250	--	--	290	--	--
Benzo[a]pyrene	325	--	--	310	--	--
Chrysene	280	480, 505, 560	6.0	272	515, 550	0.45
Coronene	300	515, 550	15.3	310	525, 560	1.1
1,2,5,6-Dibenzanthracene	300	565	19.2	300	500, 555	8.4
1,2,4,5-Dibenzpyrene	334	--	--	280	--	--
Fluoranthene	363	560	19.2	284	500, 555	1.7
Perylene	332	--	--	310	--	--
Phenanthrene	290	480, 505	6.9	260	505	0.9
Pyrene	336	--	--	322	590	2.7
Rubrene	316	--	--	286	--	--
Acridine	--	--	--	290	--	--
1,2-Benzanthracene	--	--	--	290	515	--
7,12-Dimethylbenzan-thracene	--	--	--	320	--	--
2,3,6,7-Dibenzan-thracene	--	--	--	290	--	--
1,2,7,8-Dibenzphenan-threne	--	--	--	290	510, 540	--
1,2-Benzpyrene (Benzo[e]pyrene)	--	--	--	348	420, 470, 548	--
Naphthalene	--	--	--	340	--	--
Triphenylene	--	--	--	268	395, 465, 490	--

[a] The underlined wavelengths represent the peak of greatest intensity and those used for the limits of detection.
[b] LOD = limit of detection; defined as the concentration of substance analyzed resulting in a signal-to-noise ratio of 3. The estimations were for 3-μl samples.

Reprinted with permission from Ref. 9.

determine specifically a given phosphorescent pesticide in the presence
of other weakly phosphorescent pesticides. Later, Aaron and
Winefordner (21) used sodium iodide as an external heavy-atom perturber to induce RTP from a variety of aromatic pesticides. Except
for organochlorinated compounds, sodium iodide increased signifi-
cantly the RTP of the pesticides. The absolute limits of detection
of naphthalene-like pesticides were in the nanogram range. The authors emphasized the fact that the selectivity of the RTP technique
allows one to determine specifically the nanogram amounts of naphthalene-like pesticides in the presence of larger amounts of organochlorinated pesticides.

Vo-Dinh and Gammage (22) used synchronous RTP spectrometry
to differentiate the isomers 1,2,5,6-dibenzanthracene and 1,2,3,4-
dibenzanthracene and to identify pyrene in a Synthoil extract.
Vo-Dinh (23) has considered more generally the application of synchronous luminescence spectrometry to multicomponent analysis.

TABLE 7.3 Summary of the Effect of Different Heavy Atoms on the
RTP of PAH

Compound	λ_{ex} (nm)	λ_{em} (nm)	I^a_{Pb}/I_{Tl}	I^a_{Hg}/I_{Tl}	I^a_{Ag}/I_{Tl}
Crysene	272	515	0.071	0.011	0.29
Coronene	310	525	0.054	0.051	0.44
1,2,5,6-Dibenzanthracene	300	555	0.05	0.013	0.28
Fluoranthene	284	550	0.17	0.019	--
Phenanthrene	260	505	0.046	--	0.23
Pyrene	322	590	0.067	0.015	0.16

[a] I/I_{Tl} represents the ratio of the phosphorescence signal intensity
with a heavy atom (Pb, Hg, or Ag) to the signal intensity obtained
in the presence of Tl. All solutions used were of the same concentration.

Reprinted with permission from Ref. 9.

With this approach both the excitation and emission monochromators are varied simultaneously while keeping a constant wavelength interval $\Delta\lambda$ between them. The approach has not often been employed in analysis, but it shows great promise for characterization of complex mixtures.

Von Wandruszka and Hurtubise (6) developed a method for determining PABA without separation in multicomponent vitamin tablets. The reflected RTP of PABA absorbed on sodium acetate was measured in the quantitation step. The method was rapid, very selective, and sensitive for PABA. The limit of detection was about 0.5 ng per sample spot and the standard deviation was 2.4% at 5 mg PABA per tablet. A stock PABA-containing solution was prepared from vitamin tablets by grinding 20 tablets with a mortar and pestle and then extracting with ethanol for 1 hr. The extract was filtered and the filtrate was diluted to an appropriate volume with ethanol. The diluted solution was used to prepare all other extract solutions used in subsequent PABA determinations. A list of tablet ingredients is given in Table 7.4. Von Wandruszka and Hurtubise performed interference studies and found the volume of tablet extract employed in the procedure was important for accurate determinations. However, this did not present any problems as the sensitivity of the method allowed for the determination of extremely small amounts of PABA with good precision. Their method illustrates the sensitivity and selectivity of RTP to a "real" sample.

Von Wandruszka and Hurtubise (8) indicated that benzocaine (ethyl-p-aminobenzoate) did not show room temperature phosphorescence when adsorbed on sodium acetate. However, benzocaine was hydrolyzed in dilute hydrochloric acid and was subsequently neutralized with aqueous alkali. Small amounts of this solution, when applied to sodium acetate and evaporated to dryness, showed the characteristic PABA phosphorescence on excitation with an ultraviolet lamp. They gathered no quantitative data for benzocaine, but it appears there would be little difficulty in developing a sensitive method for its determination.

TABLE 7.4   Tablet Ingredients

Ingredient	Quantity per six tablets (as labeled)
Vitamin B_1	5 mg
Vitamin B_2	5 mg
Vitamin B_6	5 mg
Folic acid	0.1 mg
PABA	30 mg
Pantothenic acid	100 mg
Niacinamide	50 mg
Inositol	1000 mg
Choline bitartrate	1000 mg
Vitamin B_{12}	15 µg
Biotin	25 µg

Reprinted with permission from R. M. A. von Wandruszka and R. J. Hurtubise, *Anal. Chem., 48*, 1784 (1976). Copyright by the American Chemical Society.

Ford and Hurtubise (4) showed that the phthalic acid isomers and other compounds exhibited room temperature phosphorescence when adsorbed on dried silica gel. They emphasized that a simple method for determining terephthalic acid in the presence of its isomers could be developed. The method would involve the separation of the dicarboxylic acids on a silica gel chromatoplate and a subsequent RTP measurement directly from the plate. Ford and Hurtubise used an alcohol/water/ammonium hydroxide (100:12:16) mobile phase described by Braun and Geenen (24) to separate the phthalic acid isomers on a silica gel chromatoplate. They observed three RTP signals which corresponded to the phthalic acid isomers. The R_f values obtained for terephthalic acid, isophthalic acid, and phthalic acid were 0.42, 0.35, and 0.18, respectively.

Ford and Hurtubise (5) obtained useful RTP in the nanogram range from several nitrogen heterocycles adsorbed on silica gel and filter paper. They found that compounds spotted from 0.1 M HCl ethanol

solutions gave very strong RTP signals. Table 7.5 gives relative
RTP intensities for the nitrogen heterocycles adsorbed on silica
gel and filter paper measured under air, nitrogen, and helium. The
data indicate that the strongest signals will be obtained with helium.
Also, filter paper yielded stronger RTP signals than silica gel
and it appeared that with dried silica gel the phosphorescence was
generally more subject to oxygen quenching than the RTP of the same
compounds adsorbed on dried filter paper. Table 7.6 shows the RTP
analytical data for the nitrogen heterocycles and PABA adsorbed
on dried silica gel. The RTP signals were measured in air without
any special drying during the measurement step. The authors emphasized that several of the compounds in Table 7.6 exhibited visible
fluorescence on dried silica gel. This additional analytical para-
meter is advantageous. For example, the identification of unknowns
is aided by fluorescence excitation and emission spectra, RTP emission spectra, and the RTP lifetime of compounds that can be measured
easily. Also, one is not restricted to a single luminescence process.
If, for example, benzo[f]quinoline (B[f]Q) were being determined
and the spot containing B[f]Q was found to contain a fluorescent
impurity but no phosphorescent impurity, RTP could be employed to
quantitate the B[f]Q in the sample without inference from the fluorescent impurity. Ford and Hurtubise stressed the use of silica
gel over filter paper in RTP work because RTP can be combined with
the speed and versatility of thin-layer chromatography to yield
a very useful analytical approach for determining nitrogen heterocycles. The authors commented that silica gel or filter paper can
be used in conjunction with high-performance liquid chromatography
(HPLC) by collecting a fraction from a HPLC column, adsorbing the
components on the solid surface, and then measuring either fluorescence or RTP. Ford and Hurtubise (25) used this approach for the
identification of B[f]Q and phenanthridine in a shale oil sample.
They separated both compounds by a combination of open column chromatography and HPLC. RTP excitation and emission spectra, fluorescence emission spectra, and chromatographic data were used for identification of the compounds.

TABLE 7.5 A Comparison of the RTP Intensities of N-Heterocycles on Silica Gel and Filter Paper Measured Under Air, N_2, and He

Compound	Silica (acid)[a,b]			Filter paper (acid)[a,b]		
	Air	Nitrogen	Helium	Air	Nitrogen	Helium
Benzo[f]quinoline	55.2	67.5	84.5	100	111	116
Phenanthridine	42.4	47.8	57.5	46.0	52.0	55.0
4-Azafluorene	22.8	26.8	33.6	42.0	50.5	54.5
1,10-Phenanthroline · H_2O	7.5	10.0	12.5	17.5	19.0	22.0
4,7-Diphenyl-1,10-phenanthroline	11.3	15.5	16.5	23.5	23.7	26.0
2,9-Dimethyl-4,7-diphenyl-1,10-phenanthroline	12.0	14.3	15.3	17.5	20.0	22.5
Quinoline	7.0	9.5	10.5	20.0	22.2	23.5
Isoquinoline	8.0	8.5	10.8	25.0	28.0	28.0

[a] Each compound at 100 ng/spot.

[b] Compounds spotted from ethanol solutions which were 0.1 M HCl.

TABLE 7.6 The RTP Analytical Data from Compounds Adsorbed on Dried Silica Gel[a]

Compound	λ_{ex} (nm)	λ_{em}[b] (nm)	Linear range (ng)	LOD (ng/spot)
Benzo[f]quinoline	370	510	0-125	3
Phenanthridine	360	510	0-160	6
4-Azafluorene	330	465	0-100	10
1,10-Phenanthroline · H_2O	325	520	0-140	25
4,7-Diphenyl-1,10-phenanthroline	310	518	0-175	12
2,9-Dimethyl-4,7-diphenyl-1,10-phenanthroline	330	518	0-150	20
Quinoline	320	510	0-100	25
Isoquinoline	330	520	0-250	22
PABA	287	432	0-150	10
2,5-Pyridine dicarboxylic acid	289	435	0-120	15

[a] All compounds spotted from ethanol solutions which were 0.1 M HCl, except PABA and 2,5-pyridine dicarboxylic acid. These compounds were spotted from ethanol solutions.

[b] λ_{ex} and λ_{em} were maximized for the Schoeffel unit.

Reprinted with permission from C. D. Ford and R. J. Hurtubise, Anal. Chem., 51, 659 (1979). Copyright by the American Chemical Society.

McHale and Seybold (26) presented RTP experiments using adsorbed dyes for educational purposes. Advantages of RTP are that cryogenic equipment is not needed, expensive quartz cuvets are not needed, and students can observe and investigate easily both fluorescence and phosphorescence at room temperature.

The applications in this chapter illustrate the analytical potential for RTP. Because of their speed, simplicity, sensitivity, and selectivity, RTP applications should appear in a variety of areas such as energy, environmental, forensic, industrial, pharmaceutical, clinical, and medical research. Also, RTP should be readily adaptable to routine determinations in quality control laboratories because the RTP data can be obtained rapidly and inexpensively.

REFERENCES

1. M. Roth, *J. Chromatogr.*, *30*, 276 (1967).

2. T. Vo-Dinh, G. L. Walden, and J. D. Winefordner, *Anal. Chem.*, *49*, 1126 (1977).

3. E. M. Schulman and C. Walling, *J. Phys. Chem.*, *77*, 902 (1973).

4. C. D. Ford and R. J. Hurtubise, *Anal. Chem.*, *50*, 610 (1978).

5. C. D. Ford and R. J. Hurtubise, *Anal. Chem.*, *51*, 659 (1979).

6. R. M. A. von Wandruszka and R. J. Hurtubise, *Anal. Chem.*, *48*, 1784 (1976).

7. R. M. A. von Wandruszka and R. J. Hurtubise, *Anal. Chem.*, *49*, 2164 (1977).

8. R. M. A. von Wandruszka and R. J. Hurtubise, *Anal. Chim. Acta*, *93*, 331 (1977).

9. E. L. Yen-Bower and J. D. Winefordner, *Anal. Chim. Acta*, *102*, 1 (1978).

10. E. M. Schulman and R. T. Parker, *J. Phys. Chem.*, *81*, 1932 (1977).

11. R. A. Paynter, S. L. Wellons, and J. D. Winefordner, *Anal. Chem.*, *46*, 736 (1974).

12. T. Vo-Dinh and J. D. Winefordner, *Appl. Spectrosc. Rev.*, *13*, 261 (1977).

13. S. L. Wellons, R. A. Paynter, and J. D. Winefordner, *Spectrochim. Acta*, Part A, *30*, 2133 (1974).

14. T. Vo-Dinh, E. L. Yen, and J. D. Winefordner, *Anal. Chem.*, *48*, 1186 (1976).

15. E. L. Yen-Bower and J. D. Winefordner, *Anal. Chim. Acta, 101,* 319 (1978).

16. I. M. Jakovljevlc, *Anal. Chem., 49,* 2048 (1977).

17. C. G. de Lima and E. M. de M. Nicola, *Anal. Chem., 50,* 1658 (1978).

18. P. G. Seybold and W. White, *Anal. Chem., 47,* 1199 (1975).

19. T. Vo-Dinh, E. L. Yen, and J. D. Winefordner, *Talanta, 24,* 146 (1977).

20. J. J. Aaron, E. Kaleel, and J. D. Winefordner, *J. Agr. Food Chem.,* in press.

21. J. J. Aaron and J. D. Winefordner, *Analysis, 7,* 168 (1978).

22. T. Vo-Dinh and R. B. Gammage, *Anal. Chem., 50,* 2054 (1978).

23. T. Vo-Dinh, *Anal. Chem., 50,* 396 (1978).

24. D. Braun and H. Geenen, *J. Chromatogr., 7,* 56 (1962).

25. C. D. Ford and R. J. Hurtubise, *Anal. Lett., 13(A6),* 485 (1980).

26. J. L. McHale and P. G. Seybold, *J. Chem. Educ., 53,* 654 (1976).

8

Applications in Environmental Research

Solid surface luminescence analysis has found extensive application
in environmental research, particularly air pollution research.
Methods have been developed for quantitative and qualitative analysis
of a variety of pollutants. Generally, the pollutants are separated
by thin-layer chromatography (TLC) and the fluorescence of the separated components is measured directly from the solid surface. Room
temperature phosphorescence (RTP) has been applied to only a few
"real" samples but certainly many more applications will appear
in the future. The previous chapter considered the RTP properties
of some potential pollutants, namely, polycyclic aromatic hydrocarbons
and nitrogen heterocycles. In this chapter, some selected analytical
examples will be considered to show the wide applicability of solid
surface luminescence analysis to environmental research.

AIR POLLUTION

In the 1960s, Sawicki and co-workers pioneered in the development
and application of solid surface luminescence analysis in air pollution research. Sawicki (1) has reviewed fluorescence analysis
in air pollution research, and Sawicki and Sawicki (2) have reviewed
the role of thin-layer chromatography in air pollution research
with the luminescence characterization, identification, and quantitation emphasized. Several examples of Sawicki's work will be

considered, and the reader is referred to the above reviews for
additional examples.

Sawicki, Stanley, and Elbert (3) used an Aminco-Bowman spectrofluorometer with an Aminco automatic scanning attachment to examine
fluorometrically both paper and thin-layer chromatograms. To demonstrate the characterization of nitrogen heterocycles, they separated
a standard mixture of five nitrogen heterocycles with a cellulose
stationary phase and a DMF/H_2O (35:65) mobile phase. The compounds
separated were dibenz[a,h]acridine, dibenz[a,j]acridine, 12-methybenz-
[a]acridine, benzo[l,m,n]phenanthridine, and acridine. After separation, the chromatoplate was treated with trifluoroacetic acid
fumes and then scanned directly for fluorescence data. By selection of appropriate excitation and emission wavelengths the compounds
were differentiated and characterized readily. For example, at
an excitation wavelength of 402 nm and an emission wavelength of
440 nm only the dibenzacridines gave fluorescence bands. In addition, the emission spectra and excitation spectra of the compounds
were obtained directly from the chromatoplate. R_f values were also
used in characterization, and all the information was obtained directly
from the chromatoplates. A similar approach was used to characterize
a standard mixture of polycyclic aromatic hydrocarbons. Benzo[a]-
pyrene, benzo[e]pyrene, benzo[k]fluoranthene, and perylene were
separated on cellulose acetate with ethanol/toluene/water (17:4:4).
The authors also showed a linear relationship could be obtained
between the area under the fluorescence band and the amount of
benzo[a]pyrene on the cellulose acetate chromatoplate. Sawicki
and co-workers successfully applied solid surface fluorescence analysis
to particulates obtained from air and from air pollution source
effluents. Fluorometric data showed the presence of dibenz[a,h]-
acridine in an atmospheric sample whose main contaminant was coal-
tar-pitch fumes. Other fluorometric data from an urban atmospheric
sample indicated the presence of benzo[a]pyrene and benzo[k]fluoran-
thene. The researchers concluded their method worked well for many
compounds in amounts ranging from nanogram to microgram. Useful

excitation and emission spectra were obtained for several compounds,
and the general approach was shown to be of value in air pollution
analysis.

Sawicki et al. (4) and Sawicki, Meeker, and Morgan (5) further
discussed various characterization procedures for urban air samples
for nitrogen heterocycles and polycyclic aromatic hydrocarbons.
Generally the procedure involved extraction of a basic fraction from
air particulate matter, column chromatography of this fraction, thin-
layer chromatography of the column chromatographic fraction, and
spectrofluorometric examination of the separated components on the
chromatoplate. In the thin-layer chromatography step the chromatoplate was treated with ammonia fumes and a variety of fluorescent
spots appeared. Fluorescence excitation and emission spectra were
then obtained directly from some of the components on the chromatoplates. Nineteen nitrogen heterocycles were characterized in urban
airborne particulates. Ten of the compounds were characterized
unequivocally and the other nine were shown to be alkyl derivatives.
Their results indicated that hundreds of other basic compounds were
present, but they did not characterize these compounds. Similar
procedures were applied to polycyclic aromatic hydrocarbons.

Sawicki and Pfaff (6) introduced quenchophosphorimetry and
applied this new approach to the analysis of several compounds.
In quenchophosphorimetry the selectivity of luminescence analysis
is improved and is particularly important for investigating mixtures
of compounds. For example, with quenchophosphorimetry it is possible
to selectively quench the phosphorescence of several components in
a mixture and then obtain phosphorescence data on the component or
components of interest. Sawicki and Pfaff used glass-fiber paper,
filter paper, and thin-layer adsorbents in their work. Components
on the solid surface with and without quencher were submerged in
liquid nitrogen and then exposed to ultraviolet radiation. The
radiation was removed, and the phosphorescence of the spots in the
dark was noted quickly. Table 8.1 gives quenching results for several
compounds under a variety of conditions. In concentrated nitrotoluene

TABLE 8.1 Phosphorescence Quenching on Glass-fiber Paper[a]

Compound	Dry	Wet (EPA)[b]	CS$_2$	Me$_2$N·NH$_2$	CH$_3$NO$_2$	NO$_2$/TFA[c] (1:1)	NO$_2$/H$_2$SO$_4$ (1:1)	o-Nitrotoluene[d] 3%	o-Nitrotoluene[d] 10-30%	C$_2$(CN)$_4$[e]	TFA
2-Acetylphenan-threne	G	G	G	BG	G	BG	--	G	--	--	--
Anthraquinone	G	G	G	--	G	G	--	G	--	--	G
Benzophenone	BG	1BG	--	--	--	--	B	B	--	1Pk	1BG
Benzo[f]quinoline	1G	G	G	BG	1G	G	G	G	--	G	G
Carbazole	--	B	--	B	--	--	--	--	--	--	1B
Dibenzothiophene	pB	B	--	B	--	--	pBG	pB	--	--	B
p-Hydroxyaceto-phenone	pB	B	--	B	--	B	BG	--	--	B	BG
2-Naphthylamine	pG	pG	--	pG	--	--	--	--	--	--	--
p-Nitroaniline	YG	YG	Y	Y	Y	--	--	YG	YG	--	--
2-Nitrofluorene	G	G	YG	--	GY	Y	--	G	G	G	--
Triphenylamine	B	B	--	B	--	--	--	--	--	--	--
Triphenylene	G	B	--	B	--	pB	--	--	--	--	B
Xanthone	B	B	--	--	--	G	YG	B	--	B	G

a B = blue, G = green, l = light, p = pale, Pk = pink, Y = yellow, -- = quenched.
b EPA = ethyl ether/isopentane/ethanol (5:5:2).
c NO$_2$/TFA = nitrogen dioxide and trifluoroacetic acid.
d In toluene.
e 5% in dimethylformamide/trifluoroacetic acid (1:1).

Reprinted with permission from Ref. 6.

solution, the nitro compounds p-nitroaniline and 2-nitrofluorene
phosphoresced while the other compounds did not. In carbon disulfide,
nitromethane, and acidic tetracyanoethylene solution, the phosphorescence of most of the compounds was quenched. All the compounds in
Table 8.1 can be made phosphorescent or nonphosphorescent by changing
the solvent. The quenching phenomenon depends on the solvent, the
reagent, the pH of the solvent system, and the structure of the
test substance. It can occur through complex formation, salt formation, redox reactions, organic synthetic reactions, or through interference with the radiative processes to the ground state in a
molecule. Sawicki and Pfaff commented that phenanthrene adsorbed
on glass-fiber paper was highly phosphorescent in the presence of
ethyl ether/isopentane/ethanol (EPA) solution but was essentially
nonphosphorescent in carbon disulfide or nitromethane. Also, carbazole on glass-fiber paper was highly phosphorescent in EPA but its
phosphorescence was quenched in carbon disulfide and nitromethane
solutions. They demonstrated the use of quenchophosphorimetry by
obtaining the phosphorescence excitation and emission spectra of
anthraquinone on glass-fiber paper with carbon disulfide as solvent
in the presence of benzophenone. Normally benzophenone gives a
strong phosphorescence, but the phosphorescence was quenched in
carbon disulfide. The authors gave several other examples and concluded because of the high selectivity of quenchophosphorimetry
fewer separation steps would be required when applied to the analysis
of various mixtures. Also, the approach was complementary to fluorescence and quenchofluorometric techniques (7). RTP techniques should
have application in quenchophosphorimetry; however, no applications
have appeared in the literature.

    Sawicki and Pfaff (8) developed methods for direct analysis
of aromatic compounds on paper and thin-layer chromatograms by spectro-
phosphorimetry at liquid nitrogen temperature, and they applied

these approaches to air pollution analysis. The phosphorescence
work was done with an Aminco-Keirs spectrophosphorimeter. The separated components were cut from filter paper, glass-fiber paper, or
thin-layer chromatoplate. The solid surface was put into a quartz
sample tube, and the tube was positioned upside down in the cell
compartment so the spot to be examined was in the light beam and
perpendicular to it. Most of the compounds they investigated showed
little or no phosphorescence when present on glass-fiber paper as
dry spots. Sawicki and Pfaff emphasized that drastic changes could
occur in phosphorescence spectra for adsorbed compounds in a wet
state compared to a dry state. The solvent and pH employed were
an important consideration. Sawicki and Pfaff characterized readily
airborne particulate samples for phenanthrene and benzo[e]pyrene
by obtaining phosphorescence spectra of the compounds adsorbed on
glass-fiber paper. The compounds were separated initially by alumina
column chromatography. The nitrogen heterocycles, benzo[h]quinoline
and benzo[c]acridine were characterized in a basic fraction of coal-
tar pitch by obtaining their phosphorescence emission spectra from
glass-fiber paper. The spectra of a phosphorescent compound, its
salt, its reduced or oxidized forms, or its derivatives, could be
obtained in a range of solvents on glass-fiber paper. Detection
limits ranged from 0.1 ng to microgram amounts. The room temperature
phosphorescence approaches discussed in Chapter 7 could be combined
with the above results to expand solid surface phosphorescence analysis
considerably.

Sawicki and co-workers have developed many more innovative
techniques and analytical methods employing solid surface luminescence analysis. These approaches have been applied to air pollution
analysis; however, the methods can be applied in many other areas
of analysis. The reader is referred to the references listed at
the end of this chapter for additional information (9-21).

Stanley, Bender, and Elbert (22) have summarized the quantitative aspects of thin-layer chromatography for air pollution samples
employing fluorescence in the measurement step. They have given
important experimental details and commented that direct optical
examination of spots on a chromatoplate is one of the quickest and
simplest means of identifying and estimating complex compounds in
samples at nanogram and microgram amounts. In situ examination
of spots on a chromatoplate minimizes the laborious task and inherent
losses associated with elution of components from a chromatoplate.
Also, techniques involving removal of compounds from chromatoplates
present difficulties in obtaining reproducible data.

In some situations it is necessary to remove the fluorescent
component from the chromatoplate, extract, and then obtain the solution fluorescence spectra. Bender (23) investigated one-dimensional
and two-dimensional thin-layer chromatographic methods and fluorometric identification and estimation of dibenzo[a,e]pyrene from
urban airborne particulate matter. He found dibenzo[a,e]pyrene
placed on a chromatoplate and developed one-dimensionally gave a
higher R_f value compared to the value obtained with the same system
in a two-dimensional procedure. Thus, the two-dimensional method
involved removing each fluorescent spot, extracting, and obtaining
the fluorescence spectra of the components.

Stanley, Morgan, and Meeker (24) used glass-fiber paper impregnated with silica gel to separate 7H-benz[de]anthracen-7-one and
phenalen-1-one found in airborne particulates. Development of the
chromatoplates was completed in 15 min. Direct fluorometric measurement and scanning were performed on the separated components. The
authors commented that most air pollution data represent mean concentrations and give an incomplete picture of a city's air quality.
Diurnal concentrations are equally important. They concluded that
in their continuing efforts to assess the applicability of methods
to analysis of sequential air samples direct spectrofluorometric
examination of TLC chromatograms was the quickest way to characterize
and estimate compounds present in complex organic mixtures. The

method gave a relative standard deviation of ±3.2% and a detection limit of 5 ng for 7H-benz[de]anthracen-7-one. For phenalen-1-one the precision was ±1% and the detection limit was 2 ng.

Malý (25) searched for a more effective separation system for separating polycyclic aromatic hydrocarbons released from high-temperature tars than those discovered up until 1971. He found that Whatman No. 4 filter paper impregnated with 10% liquid paraffin as a stationary phase and methanol saturated with liquid paraffin as a mobile phase gave better separation, but 6 hr were needed for development of the impregnated paper. Malý employed fluorescence detection of the separated polycyclic aromatic hydrocarbons on the impregnated filter paper.

Tomingas, Voltmer, and Bednarik (26) discussed direct fluorometric analysis of polycyclic aromatic hydrocarbons on thin-layer chromatoplates for airborne dust samples and animal tissues. The polycyclic aromatic hydrocarbons in the samples were concentrated by Soxhlet extraction or by sublimation, further separated with Sephadex LH20, and finally separated on 30% acetylated cellulose chromatoplates. The compounds were determined directly on the chromatoplates by measuring the fluorescence of the separated compounds with a Farrand chromatogram analyzer. Detection limits were in the nanogram range and the calibration curves were linear from about 1 to 100 ng. The authors emphasized, as have others, that direct fluorescence analysis on chromatoplates requires quantities of 0.1 to 0.001 μg per compound, but for solution fluorescence analysis a 10 to 50 times larger quantity is necessary. This is one of the important advantages of solid surface luminescence analysis. The authors concluded the procedures they developed were simple and sensitive for the simultaneous determination of 12 polycyclic aromatic hydrocarbons.

Swanson et al. (27) described a procedure for the routing measurement of benzo[a]pyrene in the atmosphere that has the capacity of determining 24 samples per day with an average recovery of 98.9 ± 5% based on spiked blank filters. They used a Perkin-Elmer MPF-3

fluorescent spectrophotometer with a thin-layer plate scanning attachment, a Perkin-Elmer 048 digital integrator, and a recorder
for all measurements from thin-layer chromatoplates. Concentrated
extracts from air samples and standards were spotted on 20% acetylated
cellulose chromatoplates and developed with ethanol/methylene
chloride (2:1). The researchers screened 105 commercially available polycyclic aromatic hydrocarbons for potential interferences
with benzo[a]pyrene. They employed three characteristics of a compound to decide if it would interfere, namely, the R_f value of the
compound, and the excitation and emission characteristics of the
compound. Only anthanthrene was found to be a potential interfering
component; however, at the excitation and emission wavelengths employed for benzo[a]pyrene, anthanthrene emitted weakly. They determined that in most cases anthanthrene contributed less than a
1% error in the analysis. By judicious selection of excitation and
emission wavelengths, fluorescence from interfering components can
in many situations be eliminated or minimized. This was especially
true for the polycyclic aromatic hydrocarbons they investigated.

WATER POLLUTION

Solid surface fluorescence analysis has been applied successfully
to the detection, identification, and quantitation of polycyclic
aromatic hydrocarbons. Keegan (28) developed methods for the determination of polycyclic aromatic hydrocarbons in natural water
in the part per trillion concentration range. Keegan investigated
water samples from three New Hampshire rivers. He removed the compounds from water by either continuous or batch extraction with
n-pentane. The isolated polycyclic aromatic hydrocarbons were further separated on cellulose chromatoplates with a 50% aqueous DMF
mobile phase. The developed chromatoplate was analyzed fluorometrically with a Farrand MK-1 spectrofluorometer equipped with a TLC
scanner. Fluorescence spectra were obtained directly from the chromatoplates, and quantitation was performed by scanning across the
unknown spot at fixed excitation and emission wavelengths. The

area of the resulting peak was compared with an appropriate standard
separated on the same chromatoplate. Keegan emphasized that quantitative TLC employing in situ methods is a fast growing field of analysis, requires less time, and is usually more sensitive than elution
methods. He pointed out, as have others, that there are a few disadvantages to direct scanning techniques, namely, the possibilities
of air oxidation or photodecomposition of compounds adsorbed on
the solid support. These are not serious disadvantages in most
situations and can be avoided usually with the appropriate experimental conditions. Keegan evaluated several chromatographic adsorbents and sample application techniques. He employed cellulose solid
support, as mentioned, and for quantitative work he used a Brinkmann
plastic micrometer-buret to apply samples to the cellulose. He also
investigated the elution technique of removing the polycyclic aromatic
hydrocarbons from the chromatoplates. He obtained very low recoveries and abandoned this approach in favor of in situ evaluation.
The range of linearity for the calibration curves was generally
between 5 and 50 ng. The relative standard deviation was 1.33%
for scanning a single 50 ng benzo[a]pyrene spot. The reproducibility
of scanning 6 identical 50 ng benzo[a]pyrene spots was also studied.
After the chromatoplate was developed and dried, the benzo[a]pyrene
spots were scanned four times each in random order. The relative
standard deviation for a single spot was 2.41% and for six spots
was 10.0%. These results included errors for the entire measurement
including spot application, variation of light source, decomposition,
and instrumental error. Keegan stated that the large difference
between the errors for a single spot and all of the spots showed
that the major portion of the error was associated with spot application. The use of an automatic spotting device could reduce the
relative standard deviation to approximately 2%. Keegan's methods
were extremely sensitive, and he applied them to 14 samples of river
water. Identification of polycyclic aromatic hydrocarbons was based
on R_B values and fluorescence spectra. All identified spots were
quantitated by fluorometric TLC scanning. The methods developed

were sufficiently rapid for routine monitoring of polycyclic aromatic hydrocarbons in natural water systems.

Hurtubise, Phillip, and Skar (29) developed a method for determining benzo[a]pyrene in filtered retort water from an in situ shale oil process by employing liquid-liquid extraction, dry-column chromatography, thin-layer chromatography and fluorescence spectrometry. Benzo[a]pyrene was identified and determined by detecting its fluorescence directly from 30% acetylated cellulose chromatoplates. The accuracy and precision of the method were good, and the limit of detection was approximately 0.08 parts per billion. They did not obtain reproducible results with unfiltered retort water samples. They postulated this was due to the adsorption of benzo[a]pyrene on suspended material in unfiltered retort water and thus extraction efficiency was reduced.

Basu and Saxena (30,31) used flexible polyurethane foam plugs to concentrate trace quantities of polycyclic aromatic hydrocarbons from large volumes of finished and raw water. In part of their work, they compared TLC coupled with fluorescence detection on aluminum oxide-acetylated cellulose thin-layer chromatoplates with gas-liquid chromatography/flame ionization detection. They concluded TLC coupled with fluorometric detection was more sensitive, more selective, and had greater sample capacity than gas-liquid chromatography/flame ionization detection (30). The emission and excitation spectra for identification of polycyclic aromatic hydrocarbons were obtained directly from the chromatoplates with an Aminco thin-film scanner attached to an Aminco-Kiers spectrophotofluorometer. Quantitation was performed with the same instrument and attachment, exciting at an appropriate wavelength and measuring the resulting fluorescence. They applied their methods to polycyclic aromatic hydrocarbons in selected U.S. drinking waters and raw water sources and found fluorescence methods were capable of detecting polycyclic aromatic hydrocarbons at lower than ng/liter levels and the TLC provided excellent resolution. Additionally, very good fluorescence excitation and emission spectra of polycyclic aromatic hydrocarbons were obtained from the raw water samples.

POLYCYCLIC AROMATIC HYDROCARBON STANDARDS ON
ACETYLATED CELLULOSE

The previous two sections were concerned with the analysis of aromatic
components in air and water. Below is analytical information for
the fluorometric investigation of standard polycyclic aromatic hydrocarbons on acetylated cellulose.

In 1970 Toth (32) discussed in situ fluorometry of polycyclic
aromatic hydrocarbons separated by thin-layer chromatography on 40%
acetylated cellulose. He employed an Aminco-Bowman spectrophotofluorometer to measure fluorescence signals and to obtain fluorescence
excitation and emission spectra. By this approach he was able to
determine and identify most of the compounds. On the average, the
limit of detection was estimated to be 0.001-0.1 µg, and standard
curves showed fairly good linearity up to 0.2 µg for a particular
compound.

Woidich et al. (33) reported experiments on thin-layer chromatographic separation of 14 polycyclic aromatic hydrocarbons on 30%
acetylated cellulose using numerous mobile phases. They described
the quantitative estimation of benzo[a]pyrene at levels of 50 pg
to 10 ng using in situ fluorescence measurements with a Zeiss PMQII
chromatogram scanner. The limit of detection for benzo[a]pyrene
was about 10 pg and measuring conditions for 14 polycyclic aromatic
hydrocarbons were given.

STABILITY OF BENZO[a]PYRENE ON SILICA GEL PLATES FOR
HIGH-PERFORMANCE THIN-LAYER CHROMATOGRAPHY (HPTLC)

Seifert (34) investigated the instability of benzo[a]pyrene on high-
performance thin-layer chromatoplates. Seifert employed a Zeiss
chromatogram spectrophotometer to investigate the decrease in fluorescence intensity of benzo[a]pyrene adsorbed on the silica gel high-
performance thin-layer chromatoplates. Figure 8.1 shows the variation
in fluorescence intensity for successive scans on different spots
from different plates relative to the first scan (0%). Each scan
took about 100 sec. For curve 1 the signal is relatively constant

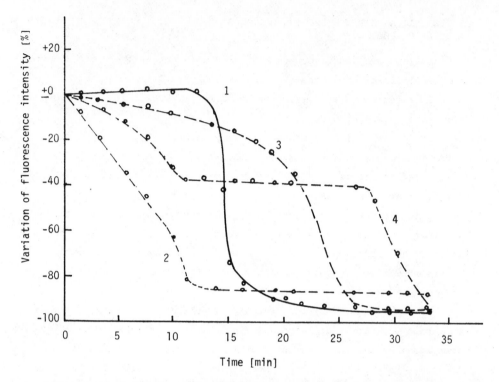

FIG. 8.1 Stability of benzo[a]pyrene on untreated HPTLC silica
gel plates. Curve 1, 1.5 ng of BaP per spot (plate A); curve 2,
3 ng of BaP per spot (plate B); curves 3 and 4, 30 ng of BaP per
spot (plate C). (Reprinted with permission from Ref. 34.)

for more than 10 min. However, for the other curves there is wide
variation in the fluorescence signals with time. Seifert concluded
in situ evaluation of benzo[a]pyrene was not possible on HPTLC silica
gel plates. He circumvented this problem by impregnating the HPTLC
silica gel plates with 5% paraffin in light petroleum (bp 60-80°C).
The stability of the fluorescence signal is illustrated in Fig.
8.2. The results show that HPTLC silica gel plates impregnated
with 5% paraffin in light petroleum have potential use for the in
situ determination of benzo[a]pyrene. However, more work is needed
to demonstrate the applicability of this approach to environmental
samples.

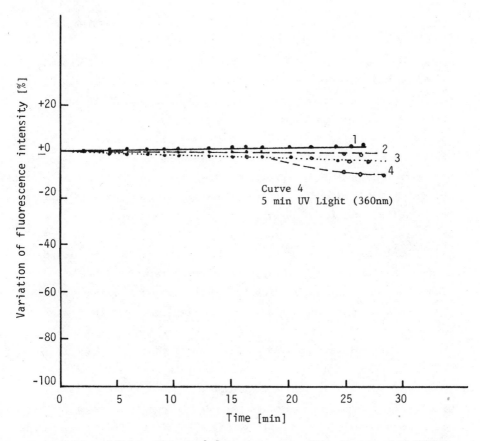

FIG. 8.2 Stability of benzo[a]pyrene on paraffin-impregnated HPTLC
silica gel plates. Curve 1, 37 ng of BaP per spot (plate A); curve
2, 1.9 ng of BaP per spot (plate B); curve 3, 1.5 ng of BaP per
spot (plate C); curve 4, 1.1 ng of BaP per spot (plate D). (Reprinted
with permission from Ref. 34.)

    The first three sections in this chapter gave several examples
of the use of acetylated cellulose as a stationary phase for separation of polycyclic aromatic hydrocarbons. Also, it has proved to
be a useful solid surface for in situ fluorescence measurements
of these compounds. It has been the author's experience that benzo[a]-
pyrene is very stable on 30% acetylated cellulose and its fluorescence characteristics remain unchanged for *several* days.

CHEMICAL SPOT TESTS FOR CERTAIN CARCINOGENS ON METAL, PAINTED,
AND CONCRETE SURFACES

Weeks, Dean, and Yasuda (35) developed procedures for the formation
of chromogenic and fluorogenic derivatives of certain aromatic amines
designated by the Occupational Safety and Health Administration
(OSHA) as cancer-suspect agents. They employed filter paper and
well-defined painted metal and concrete as standard surfaces to evaluate limits of detection for the compounds. The fluorogenic reagents
used were fluorescamine, o-phthalaldehyde, m-phthalaldehyde, and p-phthal-
aldehyde. The limit of detection values in terms of grams of analyte
per cm^2 of surface being analyzed ranged from the low-nanogram to
5-µg level depending on the compound, sampling technique, and surface. Table 8.2 shows fluorescence limits of detection for aromatic
amines visualized on filter paper. For concrete and painted surfaces, they used a leaching technique. The compound of interest
was leached from the surface with solvent-impregnated filter paper.
For example, a filter paper was placed upon the surface and five
drops of methanol were applied to the paper's center and kept in
intimate contact with the surface until the paper dried. The paper
was removed and the visualization reagent applied to the filter
paper. After a suitable reaction period, the paper was examined
for fluorescence. In some cases the researchers applied fluorescamine
directly to a concrete surface and were able to observe the fluorescence of some of the compounds in the range of 200-5000 ng cm^{-2}.
For metal surfaces, a "swipe" method was used. This involved uni-
directional wiping of approximately 10 cm^2 of the surface with filter
paper followed by the application of the visualization reagent.
They concluded that the leaching technique was more sensitive and
thus used it in later tests. However, in field survey and monitoring
work, they recommended a combination of the "swipe" and leaching
techniques. Also, by employing both chromogenic and fluorogenic
derivation techniques as complementary methods, the confidence in
the analytical results was enhanced. The authors discussed briefly
"false positive" results. Depending on the sample being tested, an
analyst will have to decide whether to run additional spot tests or
to separate the compound and then run spot tests.

TABLE 8.2 Limit of Detection Values for Aromatic Amines Visualized on Filter Paper[a] via Fluorogenic Derivatization

Compound studied	Fluorogenic visualization reagent, limit of detection values (ng/cm^2)			
	Fluorescamine	Isomeric phthalaldehydes		
		o-	m-	p-
Aniline	3	30	b	b
o-Chloroaniline	30	b	--	--
m-Chloroaniline	3	30	--	--
p-Chloroaniline	3	30	b	b
o-Toluidine	3	800	b	b
m-Toluidine	3	30	--	--
p-Toluidine	3	30	--	--
o-Tolidine	5	--	--	--
4,4'-Methylenedianiline	3	30	800	800
4,4'-Methylenebis(2-chloroaniline) (MOCA)[c]	3	800	800	800
Benzidine[c]	3	3	80	80
3,3'-Dichlorobenzidine[c]	3	800	150	80
α-Naphthylamine[c]	30	80	80	80
β-Naphthylamine[c]	3	3	80	80
4-Aminobiphenyl[c]	3	3	80	80

TABLE 8.2   (Continued)

a Whatman 42 filter paper.

b Not detected at the level of 800 ng/cm^2.

c Compounds designated as cancer-suspect agents by OSHA.

Reprinted with permission from R. W. Weeks, Jr., B. J. Dean, and S. K. Yasuda, *Anal. Chem.*, *48*, 2227 (1976). Copyright by the American Chemical Society.

PETROLEUM OIL AND COAL-TAR ANALYSIS

Killer and Amos (36) employed TLC and fluorescence detection of some components in petroleum products. They obtained a fluorescence spectrum of perylene directly from a chromatoplate. Also, Coates (37) used TLC and fluorescence detection and other detection methods for the analysis of lubricating oils and oil additives. Amos (38) has discussed the application of TLC in the heavy organic industry and included a brief discussion of direct measurement of fluorescence from chromatoplates. Matthews (39) combined TLC and fluorescence detection from developed chromatoplates to distinguish petroleum and coal-tar oils and residues. His approach was based on varying chromatographic characteristics of the samples and the resulting fluorescence of the components on the chromatoplates when viewed under ultraviolet radiation. The sample of a chloroform extract was developed on an alumina T chromatoplate with acetone as a mobile phase and then viewed under ultraviolet radiation. Usually the sample could be classified into one or more classes indicated below.

    Class I: Lubricating greases, petrolatum, lube oils
    Class II: Engineering oils, diesel fuel oils, solvent fuels,
                anthracene oil
    Class III: Coal-tar products
    Class IV: Asphalts, bitumens, crude oils, fuel oils
    Class V: Fluorescing oils not fitting standard pattern
    Class VI: Nonfluorescing oils

The choice in a given class could be further narrowed by employing
other TLC systems and fluorescence characteristics of the separated
components. Matthews commented that his separation approach could
not be used to separate complex mixtures but was effective with
simple mixtures. Nevertheless he applied his methods successfully
to water pollution control problems such as identifying the source
of pollution or minimizing the choices of the sources of pollution.
Bentz (40) reviewed oil spill identification and briefly discussed
the role of TLC and direct fluorescence detection in oil-spill work.

SHALE OIL ANALYSIS

Hurtubise et al. (41) developed methods for characterization and
identification of polycyclic aromatic hydrocarbons in shale oil.
One simple and rapid method involved developing shale oil samples
on silica gel chromatoplates with n-hexane. The fluorescence of
the distributed components was measured with a Schoeffel spectrodensitometer and each shale oil sample studied gave a unique fluorescence profile. The areas under the curves were determined by cut-
ting out the appropriate area of chart paper and then weighing the
paper. The method provided a quick semiquantitative measure of polycyclic aromatic hydrocarbon content in shale oil. They also developed
separation methods for several polycyclic aromatic hydrocarbons and
employed fluorescence emission spectra along with chromatographic
data to identify the compounds. Figure 8.3 compares the fluorescence emission spectra of a suspect benzo[a]pyrene spot from a shale
oil sample to a standard benzo[a]pyrene spot. Both spectra were
obtained directly from a 30% acetylated cellulose chromatoplate.

Hurtubise, Skar, and Poulson (42) developed a method for the
determination of benzo[a]pyrene in shale oil by solid surface fluorescence. A two-step separation method was employed to isolate benzo[a]-
pyrene from shale oil. Dry-column chromatography was used in the
first step and TLC in the second step. Benzo[a]pyrene was identified
and determined by detecting its fluorescence directly from 30% acetylated cellulose chromatoplates; as little as 0.06 ng could be

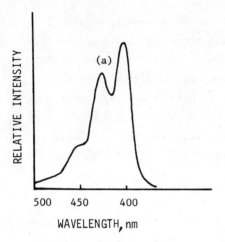
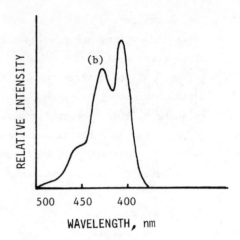

FIG. 8.3 Fluorescence emission spectra of benzo[a]pyrene spots.
(a) Suspected benzo[a]pyrene spot; (b) standard benzo[a]pyrene spot.
(Reprinted with permission from Ref. 41.)

detected. The limit of detection of benzo[a]pyrene in shale oil
was about 1.2 ppm and the reproducibility of the method was ±2.6
ppm. The percentage benzo[a]pyrene recovery in spiked shale oil
samples is given in Table 8.3. Considering the complexity of the
shale oil samples, the accuracy of the method is good over a wide
range of concentrations.

Hurtubise, Skar, and Poulson compared a Schoeffel spectrodensitometer and a Kontes densitometer for the limit of detection of

TABLE 8.3 Percentage Benzo[a]pyrene Recovery

Sample	Present (ppm BaP)	Added (ppm BaP)	Found (ppm BaP)	Recovery (%)
1	23.3	10.0	32.2	89.0
		100	117	93.7
2	8.2	10.0	17.6	94.0
		100	97.6	89.4
5	9.1	10.0	20.1	110
		100	94.2	85.1

Reprinted with permission from Ref. 42.

benzo[a]pyrene in shale oil. The limit of detection was calculated to
be 1.2 ppm (Schoeffel unit) and 2.2 ppm (Kontes unit). The Schoeffel
unit, a research spectrodensitometer with a high-intensity xenon lamp ,
source, was expected to have better performance characteristics than
the Kontes densitometer, which was designed for routine analysis. In
additional experiments the relative fluorescence intensities of benzo[a]-
pyrene standards were measured on 30% acetylated cellulose, aluminum
oxide, and silica gel chromatoplates with the Schoeffel unit. The
best limit of detection was obtained for benzo[a]pyrene on 30% acetylated cellulose chromatoplates with the fluorescence measured in
the reflection mode. The smallest amount of benzo[a]pyrene detected
in the fluorescence reflection mode on 30% acetylated cellulose
was 0.06 ng. On both silica gel and aluminum oxide the smallest
amount of benzo[a]pyrene detected was 1.6 ng.

MATRIX ISOLATION

The matrix isolation technique involves diluting the sample with
a large excess of inert gas, and then the mixture is frozen onto
a transparent window (solid surface) maintained at very low temperatures (43). The deposited solid can then be subjected to spectroscopic analysis. Stroupe et al. (44) explored the use of matrix
isolation fluorescence spectroscopy in the qualitative and quantitative analysis of polycyclic aromatic hydrocarbons. Figure 8.4
shows the matrix isolation fluorescence spectra of a chromatographic
fraction from a Synthoil sample. The spectra showed the presence
of pyrene, benz[a]anthracene, and chrysene. Later, Tokousbalides
et al. (45) reported on the analysis of the isomeric methychrysenes
by matrix isolation fluorescence and Fourier-transform infrared
spectrometry. Each of the six isomers could be identified, in mixtures containing all six compounds, by matrix isolation fluorescence
spectroscopy. The authors emphasized the complementary nature of
matrix isolation fluorescence spectroscopy and Fourier-transform
infrared spectroscopy. There have been very few analytical applications of matrix isolation luminescence spectroscopy, and this po-

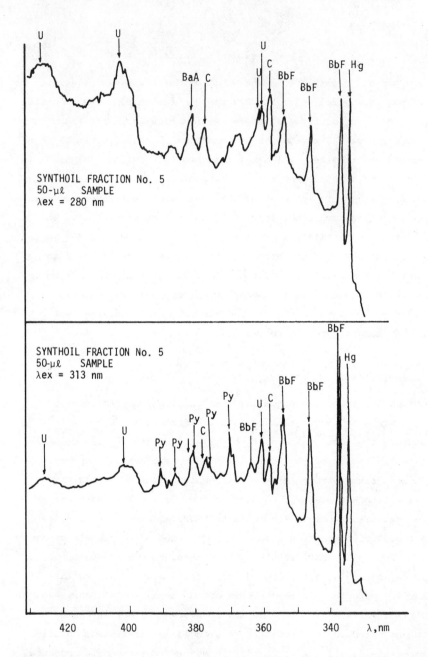

FIG. 8.4 MI fluorescence spectra of liquid chromatographic fraction from Synthoil. Identified components: BbF, benzo[b]fluorene (added as internal standard); BaA, benz[a]anthracene; C, chrysene; Py, pyrene. "U" denotes a fluorescence band which cannot at present be assigned to a specific compound; "Hg" denotes stray light (a mercury line emitted by xenon-mercury arc source). (Reprinted with permission from R. C. Stroupe, P. Tokousbalides, R. B. Dickinson, Jr., E. L. Wehry, and G. Mumantov, *Anal. Chem. 49*, 701 (1977). Copyright by the American Chemical Society.)

tentially important technique awaits further developments. Wehry
and Mamantov (46) recently reviewed the analytical instrumentation
for matrix isolation spectroscopy.

REFERENCES

1. E. Sawicki, *Talanta, 16*, 1231 (1969).

2. C. R. Sawicki and E. Sawicki, *Progress in Thin-Layer Chroma-
    tography and Related Methods*, Vol. III, A. Niederwieser and
    G. Pataki, eds., Ann Arbor Science Publishers Inc., Ann Arbor,
    Mich., 1972, Chap. 6.

3. E. Sawicki, T. W. Stanley, and W. C. Elbert, *J. Chromatogr.,
    20*, 348 (1965).

4. E. Sawicki, S. P. McPherson, T. W. Stanley, J. Meeker, and
    W. C. Elbert, *Int. J. Air Wat. Poll., 9*, 515 (1965).

5. E. Sawicki, J. E. Meeker, and M. J. Morgan, *Int. J. Air Wat.
    Poll., 9*, 291 (1965).

6. E. Sawicki and J. D. Pfaff, *Mikrochim. Acta*, 322 (1966).

7. E. Sawicki, T. W. Stanley, and W. C. Elbert, *Talanta, 11*, 1433
    (1964).

8. E. Sawicki and J. D. Pfaff, *Anal. Chim. Acta, 32*, 521 (1965).

9. E. Sawicki, T. W. Stanley, W. C. Elbert, and J. D. Pfaff, *Anal.
    Chem., 36*, 497 (1964).

10. E. Sawicki, T. W. Stanley, and H. Johnson, *Microchem. J.,
    8*, 257 (1964).

11. E. Sawicki, T. W. Stanley, J. D. Pfaff, and W. C. Elbert, *Anal.
    Chim. Acta, 31*, 359 (1964).

12. E. Sawicki, T. W. Stanley, and H. Johnson, *Mikrochim. Acta,*
    178 (1965).

13. E. Sawicki, W. C. Elbert, and T. W. Stanley, *J. Chromatogr.,
    17*, 120 (1965).

14. E. Sawicki and H. Johnson, *Microchem. J., 8*, 85 (1964).

15. J. D. Pfaff and E. Sawicki, *Chemist-Analyst, 54*, 30 (1965).

16. E. Sawicki, T. W. Stanley, W. C. Elbert, and M. Morgan, *Talanta,
    12*, 605 (1965).

17. E. Sawicki and H. Johnson, *J. Chromatogr., 23*, 142 (1966).

18. E. Sawicki and J. D. Pfaff, *Microchem. J., 12*, 7 (1967).

19. E. Sawicki, M. Guyer, and C. R. Engel, *J. Chromatogr., 30*,
    522 (1967).

20. C. R. Engel and E. Sawicki, *J. Chromatogr.*, *31*, 109 (1967).

21. E. Sawicki, M. Guyer, R. Schumacker, W. C. Elbert, and C. R. Engel, *Mikrochim. Acta*, 1025 (1968).

22. T. W. Stanley, D. F. Bender, and W. C. Elbert, *Quantitative Thin Layer Chromatography*, J. C. Touchstone, ed., John Wiley & Sons, Inc., New York, 1973, Chap. 16.

23. D. F. Bender, *Environ. Sci. Technol.*, *2*, 204 (1968).

24. T. W. Stanley, M. J. Morgan, and J. E. Meeker, *Environ. Sci. Technol.*, *3*, 1198 (1969).

25. E. Malý, *Mikrochim. Acta, 429* (1977).

26. R. Tomingas, G. Voltmer, and R. Bednarik, *Sci. Total Environ.*, *7*, 261 (1977).

27. D. Swanson, C. Morris, R. Hodgecoke, R. Jungers, R. Thompson, and J. E. Bumgarner, "Trends in Fluorescence," *1*, 22 (1978) (Perkin-Elmer Corporation, Main Avenue-M.S. 12, Norwalk, Conn.).

28. R. Keegan, Ph.D. Thesis, University of New Hampshire, Durham, New Hampshire, 1971.

29. R. J. Hurtubise, J. D. Phillip, and G. T. Skar, *Anal. Chim. Acta, 101*, 333 (1978).

30. D. K. Basu and J. Saxena, *Environ. Sci. Technol.*, *7*, 791 (1978).

31. D. K. Basu and J. Saxena, *Environ. Sci. Technol.*, *7*, 795 (1978).

32. L. Toth, *J. Chromatogr.*, *50*, 72 (1970).

33. H. Woidich, W. Pfannhauser, G. Blaicher, and K. Tiefenbacker, *Chromatographia, 10*, 140 (1977).

34. B. Seifert, *J. Chromatogr.*, *131*, 417 (1977).

35. R. W. Weeks, Jr., B. J. Dean, and S. K. Yasuda, *Anal. Chem.*, *48*, 2227 (1976).

36. F. C. A. Killer and R. Amos, *J. Inst. Petrol.*, *52*, 315 (1966).

37. J. P. Coates, *J. Inst. Petrol.*, *57*, 209 (1971).

38. R. Amos, *Talanta, 20*, 1231 (1973).

39. P. J. Matthews, *J. Appl. Chem.*, *20*, 87 (1970).

40. A. P. Bentz, *Anal. Chem.*, *48*, 454A (1976).

41. R. J. Hurtubise, J. F. Schabron, J. D. Feaster, D. H. Therkildsen, and R. E. Poulson, *Anal. Chim. Acta, 89*, 377 (1977).

42. R. J. Hurtubise, G. T. Skar, and R. E. Poulson, *Anal. Chim. Acta, 97*, 13 (1978).

43. J. S. Skirk and A. M. Bass, *Anal. Chem.*, *41*, 103A (1969).

44. R. C. Stroupe, P. Tokousbalides, R. B. Dickinson, Jr., E. L. Wehry, and G. Mamantov, *Anal. Chem.*, *49*, 701 (1977).

45.   P. Tokousbalides, E. Ray Hinton, Jr., R. B. Dickinson, Jr.,
      P. V. Bilotta, E. L. Wehry, and G. Mamantov, *Anal. Chem.*, *50*,
      1189 (1978).

46.   E. L. Wehry and G. Mamantov, *Anal. Chem.*, *51*, 643A (1979).

9

Applications in Forensic Science

Recently Gibson (1) reviewed the applications of luminescence in
forensic science. He cited 111 references and considered both solid
surface and solution luminescence applications. He discussed analysis
of phenylethylamines, morphine and its derivatives, cannabis, and
other significant drugs in forensic science. Also, he considered
characterization of materials such as fibers, hair, blood stains,
and oil spills. Solid surface luminescence has played a prominent
role in the analysis and characterization of the materials mentioned
above. Gibson stated that a major field in which luminescence has
been used is questioned-document work. Water marks can be faked
by waxing the document to make it more translucent, and thus it
will show a different luminescence from the paper. Also, most papers
possess their own luminescence that would be altered by chemical
erasure. Many fluorescent powders have been developed to show mechanical erasure alterations, and much work has gone into characterizing the luminescence of inks. Those inks that luminesce do so mainly
in the infrared region. Apparently there is a high probability
that any additions made to a document using inks of similar appearance
may be detected by a difference in infrared luminescence. Gibson
gave several additional examples of the use of solution and solid
surface luminescence analysis in forensic science. He had covered
the literature through 1976, and stated a few examples from 1977.

Below are several examples from the recent literature to illustrate the expanding use of solid surface luminescence analysis in forensic science.

Lloyd (2,3) showed that when a tire rests on a flexible strip of thin-layer chromatography (TLC) adsorbent, either of aluminum oxide or of silica gel, a fluorescent print appears within 1 hr. When excited by 366-nm mercury vapor, the fluorescence can be analyzed spectrofluorimetrically directly from the surface. Alternately, one can extract the adsorbent with cyclohexane and obtain solution fluorescence (2). Lloyd (2) considered the origins of tire fluorescence, aging effects on tread prints, direct extraction of fluorescent material from rubber, variation between tire treads, sidewall rubbers, eraser rubbers, footwear, and oil-contaminated rubbers. He employed solution fluorescence but certainly direct fluorescence measurement of separated components on TLC chromatoplates could be employed effectively. Generally three types of fluorescent material are present in the tire marks. These are various mineral oils, polynuclear aromatic hydrocarbons, and antioxidants. Usually most of the fluorescence is due to the oils, and polycyclic aromatic hydrocarbons contribute little to the fluorescence from new tires. After increasing use, the distinctive fluorescence of polycyclic aromatic hydrocarbons appears, and with a given manufacturer an excellent correlation exists between the fluorescence and tread depth.

Lloyd (3) developed a technique for enhancing the fluorescence of marks on surfaces. He found that if an organic solvent was poured onto weakly fluorescent marks, the fluorescent material hidden in rubber particles and within the surface itself was extracted rapidly and an intense fluorescence pattern, reproducing the tire pattern, appeared. The pattern rapidly disappeared as the solution diffused over and into the surface. Lloyd preserved the intensified print by spraying an approximately 15% solution of polyvinyl acetate in ethyl acetate onto the surface. A permanent record was obtained by photographing the fluorescing surface. This general approach was performed with tire marks on concrete, brick, a polythene bag,

and other surfaces. Lloyd commented that under ideal conditions
each tire could leave a distinctive print characterizing the manufacturer, and if any irregularities are produced it would be possible
to identify the tires specifically. However, ideal conditions are
rarely achieved at the scene of an accident or crime, but very useful information could be obtained from this unique solid surface
fluorescence method. Additionally, Lloyd poured liquid nitrogen
over tire marks on surfaces and brilliant phosphorescence was visible
after the ultraviolet source was removed. This approach needs further
development, but appears to have substantial analytical potential
in forensic science.

Dalrymple, Duff, and Menzel (4) utilized lasers in fingerprint
detection. Their approach consisted of laser illumination of the
exhibit under investigation, and then either direct viewing or photographing of the laser-induced fingerprint luminescence. They did
not distinguish between fluorescence and phosphorescence and may have
detected some room temperature phosphorescence. The exhibit under
investigation was illuminated with the 514.5-nm line of a continuous-
wave argon-ion laser. The laser beam illuminated an area of about
65 cm^2. They found the laser-induced fingerprint luminescence was
amenable to color photography. Fingerprints were detected on white
paper even after baking the paper at 75°C for two weeks followed
by soaking for 5 min under running water. Also, they detected fingerprints from a stainless steel knife blade, a styrofoam cup, a brown
glass bottle, living human skin, and a 14-month-old fingerprint
on a letter.

It may be possible to date fingerprints with the approach of
Dalrymple, Duff, and Menzel. For example, if one or more of the
components that give fluorescence were unstable, then relative intensities of the emission might be used to date fingerprints. They
found that old fingerprints take on an orange color, and fresh fingerprints exhibit a green color.

Later, Duff and Menzel (5) initiated a study of fingerprint-age
determination. They described a thin-layer chromatography (TLC)

study of fingerprint residue of luminescent components in palmar
sweat, and coupled this with laser excitation of luminescence.
Fingerprint material was collected by wiping fingers and palms with
a cotton swab soaked in diethyl ether and extracting the residue
into ether solution. This solution was concentrated to about 1
ml and was streaked onto a silica gel chromatoplate and developed
in diethyl ether or acetone. When the chromatogram was illuminated
by argon-laser light and viewed through laser safety goggles, six
luminescent bands were seen. These bands consisted of a strong
yellow band at the origin, an orange band at R_f = 0.31, a strong
green band at R_f = 0.78, and three weak green bands at R_f = 0.16,
0.91, and 0.98. In most cases, illumination of the chromatoplate
with an ultraviolet handlamp showed no luminescence. Exposure of
a developed chromatoplate for about 1 min to ultraviolet radiation
from a low-pressure mercury lamp caused five new luminescent green
bands to become visible on subsequent laser examination.

Duff and Menzel (5) obtained the luminescence spectrum of a
single fingerprint deposited on a glass slide. Luminescence background subtraction was performed by illuminating bare glass and sub-
tracting the resulting signal from the fingerprint luminescent bands.
Duff and Menzel did not comment whether they were measuring fluorescence, or phosphorescence, or both. However, from the instrumentation described they were probably measuring fluorescence, and any
phosphorescence that appeared at room temperature. One major aim
of their study was to characterize the key luminescent components
in fingerprint residue. They used excitation and emission spectra
for this, some obtained directly from chromatoplates, and concluded
that the green-luminescing bands were riboflavin derivatives, and
the new bands appearing after exposure to ultraviolet radiation
might be attributable to lumiflavine. The yellow band was considered
to be from a B-vitamin. They also stated their laser illumination
method represented a sensitive TLC detection method, and their approach could be extended to a greater range of materials.

Thornton (6) experimented with coumarin-6 mixtures to find
a mixture that would possess superior qualities as a fingerprint

dusting powder. One important property is the adhering qualities
of the powder. He found that by mixing an ethanolic solution of
coumarin-6 with conventional black fingerprint powder at a ratio
of 1 part of dye to 100 parts of fingerprint powder, and then allowing the ethanol to evaporate, a very useful mixture resulted. He
gave examples of fingerprints developed on a styrofoam cup and on
an apple with black powder tagged with coumarin-6 laser dye. An
argon laser was employed to excite the luminescence. Thornton stated
that only an exceedingly small amount of the powder would be needed
to adhere to the ridges of the latent fingerprint.

Lloyd (7) was able to characterize fluorescent brighteners
derived from detergents or introduced during manufacture in small
samples of fiber down to 5 μg with TLC and fluorescence detection.
Brightened, undyed fibers fluoresced and dyed fibers did not because
of quenching and self-absorption. Lloyd described special micro-
techniques for extracting the brighteners and spotting the resulting
extract onto silica gel chromatoplates under photographically safe
light. The developed chromatoplates were examined in ultraviolet
radiation (366 nm) and the positions of the separated compounds
were correlated with standard compounds. Lloyd also investigated
a number of washing powders for fluorescent brighteners. He considered extraction conditions, effects of ultraviolet light, and
chromatographic conditions. He examined samples of fiber for brighteners from 61 articles differing in origin. No brighteners were
found in 12 of the samples. He was able to correlate several of
the remaining samples with various washing powders containing brighteners that were examined by similar techniques. Lloyd concluded
that the techniques described would probably find most use in the
correlation of transferred fibers with possible points of origin,
and this type of comparison would be enhanced by high-performance
thin-layer chromatography (HPTLC) and liquid-column chromatography.

Calloway and Jones (8) investigated the discrimination of 143
glass samples by low-temperature phosphorescence analysis. Most of
the samples were window glass. The phosphorescence was emitted
by trace impurities in the glass. In future work, Calloway and

Jones plan to compare the results of their phosphorescence analysis
with trace-element content in glass samples. They constructed a
cold-finger Dewar flask to maintain the glass samples near liquid-
nitrogen temperature during the analyses. The glass chips were mounted
in indium. A modified Aminco-Bowman spectrophotofluorometer was
used to record all of the glass phosphorescence data. They also
employed a phosphoroscope to exclude scattered radiation and fluorescence. The phosphorescence emission spectra showed two major bands
from approximately 500 to 600 nm and 630 to 800 nm. They concluded
that there were at least four types of phosphorescing species in the
glass samples. Excitation at 385 nm excited species with phosphorescence that peaked at 725 nm, and excitation at 225 nm and 425 nm pref-
erentially excited species with phosphorescence that peaked at 540 nm
and 725 nm. Excitation at 325 nm excited species with phosphorescence bands that peaked at 575 nm and 740 nm. Calloway and Jones
pointed out that most crime laboratories use refractive index or
density of glasses for comparison of samples. As a result of modern
glass production techniques, glass formulations are closely controlled
from batch to batch. Because of greater uniformity of window
glass, the value of analytical techniques that rely on differences
in refractive index and density is questionable. In their work,
most of the 21 glass samples that had the same refractive index were
found to have different phosphorescence spectra. They decided in
their initial study only to measure the phosphorescence intensity
ratio of the samples at 740 nm and 550 nm. For each of the 21 glass
samples, the phosphorescence spectra were recorded once at six different excitation wavelengths. The ratio I_{740}/I_{550} was measured
10 times, with the sample removed and remounted or changed between
each measurement. At least eight different chips were used in these
experiments. The ratio of red to green phosphorescence excited
at 425 nm for the 21 glass samples with similar refractive indexes
is shown in Fig. 9.1. It can be seen that the phosphorescence ratio
technique can be employed to distinguish many of these glasses.
Samples E, F, and G were not distinguishable with 425-nm excitation
but were distinguishable when 279-nm excitation was used. When

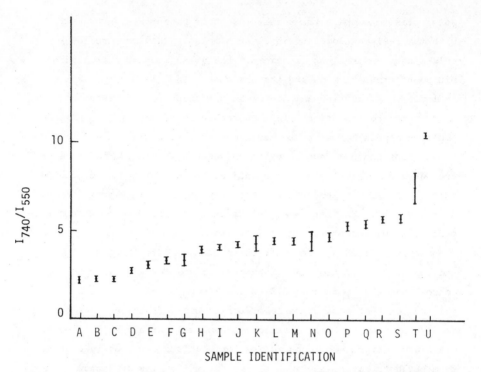

FIG. 9.1 Ratio of red to green phosphorescence, 425-nm excitation, for 21 glass samples indistinguishable by their refractive indexes. (Reprinted from Ref. 8 with permission of the American Society for Testing and Materials, 1916 Race Street, Philadelphia, Pa. 19103. Copyright.)

other excitation wavelengths were employed, it was possible to distinguish samples not distinguishable by a single excitation wavelength. For example, in Fig. 9.2 the ratio I_{740}/I_{550} for 425-nm excitation was plotted against the corresponding ratio for 270-nm excitation. Figure 9.2 indicates the group of 21 samples is separated into 16 distinguishable groups. The data for the other 4 excitation wavelengths indicated all but 2 pairs of the 21 samples were distinguishable. Calloway and Jones (8) addressed the question: Within a single window, do the phosphorescent properties vary as they do among samples from different windows? To answer this question, they measured samples from the four corners of two different windows. Within

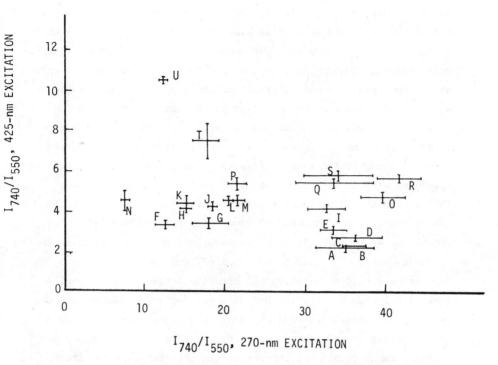

FIG. 9.2   Ratio of red to green phosphorescence, 270- and 425-nm
excitation.   (Reprinted from Ref. 8 with permission of the American
Society for Testing and Materials, 1916 Race Street, Philadelphia,
Pa. 19103.   Copyright.)

the precision of the measurements, they found that different loca-
tions within a single window have the same phosphorescent properties.
Also, they emphasized the need for additional experiments in which
a much larger number of samples would be analyzed, and the full
range of luminescent properties to establish the optimum capability
of the luminescence approach would be used.

Nowicki (9) presented data which showed the use of fluorescamine
in forensic toxicological analysis of amphetamine and differentiation
from methamphetamine.   Fluorescamine reacts with primary amines
to form intensely fluorescent substances and provides the basis
for a rapid and sensitive assay for compounds such as amino acids,
primary amines, peptides, and proteins (10).   Fluorescamine does

not react with secondary or tertiary amines. Amphetamine is a primary amine and gives an intense blue-green fluorescent product when reacted with fluorescamine. Methamphetamine does not give a fluorescent product. Nowicki developed a spot test procedure using a tile-welled spot plate in which borate buffer at pH 9, exhibit sample, and fluorescamine were mixed. A positive test was the formation of an intense blue-green fluorescent product. Frequently the blue-green product was further analyzed by TLC. About two microliters of spot test solution were applied to a silica gel chromatoplate and developed with chloroform/methanol (90:10). After development, the fluorescent products were detected with ultraviolet radiation and their R_f values compared to those of standard compounds. For removal of amphetamine in urine samples, the amphetamine was extracted from the urine, the extract concentrated, and the concentrate spotted and developed on a silica gel chromatoplate. A fluorescamine-acetone solution was sprayed on the developed chromatoplate and then oversprayed immediately with pH 9 borate buffer. The fluorescent product was detected with ultraviolet radiation, and naturally occurring compounds did not interfere with the analysis. A total of 325 drugs were tested for response to the fluorescamine reagent, and about 10% of the standard compounds tested yielded a positive fluorescamine test. From the molecular structure of the compounds it was concluded that a primary amine was necessary for a positive test. Nowicki found the migration rate of the amphetamine fluorescent derivative was unique. No standards tested gave the same migration rate. Also, the detection limit of amphetamine in urine samples on thin-layer chromatoplates was enhanced by a factor of 100 with fluorescamine compared to ninhydrin.

Vinson, Patel, and Patel (11) described a TLC-fluorescence method for detection of Δ^9-tetrahydrocannabinol (THC) in blood and serum. In the procedure, a fluorescent derivation was formed by reacting THC with 2-p-chlorosulfophenyl-3-phenylindone, which reacted with the phenolic group in THC. After extraction and cleanup of blood, serum, or plasma samples, the resulting residue was derivatized

and the derivative was spotted on a silica gel chromatoplate and
developed in methanol/water (95:5). After developing and drying
the chromatoplate, the plate was sprayed with visualization reagent
(saturated solution of sodium metal in methanol and dimethyl sulfox-
ide). While still wet, the chromatoplates were examined under ultraviolet light. The derivative appeared as a yellow-green fluorescent
spot at an R_f of 0.40-0.46. Several drugs were tested and found
not to give a positive response. The limit of detection for the
THC derivation was about 0.1 ng.

Lytle and Hedgecock (12) used luminol to induce chemiluminescence in the visualization of forensic bloodstains. Luminol solutions are unstable and must be mixed immediately before use. However,
Lytle and Hedgecock found that luminol/carbonate/water and sodium
perborate/water solutions could be stored separately for at least
eight weeks and retain full activity when combined. They employed
a hand-pump sprayer in applying luminol reagent. In the application of the reagent, they were able to detect blood in irregular
surfaces such as wood-finish paneling in the grooves even after
vigorous scrubbing. The luminol test proved to be a valuable field
test in the detection of blood because it was sensitive, could
rapidly screen large areas with a simple hand-pump sprayer, and was
reasonably specific for blood, even though it reacted with vegetable
peroxidases and with some metals and chemicals.

REFERENCES

1. E. P. Gibson, *J. Forens. Sci.*, *22*, 680 (1977).

2. J. B. F. Lloyd, *Analyst*, *100*, 82 (1975).

3. J. B. F. Lloyd, *J. Forens. Sci. Soc.*, *16*, 5 (1976).

4. B. E. Dalrymple, J. M. Duff, and E. R. Menzel, *J. Forens. Sci.*,
    22, 106 (1977).

5. J. M. Duff and E. R. Menzel, *J. Forens. Sci.*, *23*, 129 (1978).

6. J. I. Thornton, *J. Forens. Sci.*, *23*, 536 (1978).

7. J. B. F. Lloyd, *J. Forens. Sci. Soc.*, *17*, 145 (1977).

8. A. R. Calloway and P. R. Jones, *J. Forens. Sci.*, *23*, 263 (1978).

9.   H. G. Nowicki, *J. Forens. Sci.*, *21*, 154 (1976).

10. S. Udenfriend, S. Stein, P. Bohlen, W. Dairman, W. Leimgruber, and M. Weigele, *Science, 178*, 871 (1972).

11. J. A. Vinson, D. D. Patel, and A. H. Patel, *Anal. Chem.*, *49*, 163 (1977).

12. L. T. Lytle and D. G. Hedgecock, *J. Forens. Sci.*, *23*, 550 (1978).

10

Applications in Pesticide Analysis

Almost all the pesticide analysis applications in which luminescence
is detected from a solid surface involve measurement of fluorescence
from thin-layer chromatography (TLC) chromatoplates. Some work
has been done with the fluorescence measurement of pesticides from
chromatographic paper. Very recently Aaron, Kaleel, and Winefordner
(1), and Aaron and Winefordner (2) obtained useful room temperature
phosphorescence signals from several pesticides adsorbed on filter
paper. Some excellent reviews on the use of fluorescence in pesticide analysis have appeared. Lawrence and Frei (3) have reviewed
fluorometric derivatization for pesticide residue analysis, and
Frei and Lawrence (4) discussed fluorogenic labeling of carbamate
pesticides. Mallet, Belliveau, and Frei (5) reviewed in situ fluorescence spectroscopy of pesticides, and MacNeil and Frei (6) considered
quantitative TLC of pesticides. Argauer (7) covered both solution
and solid surface fluorescence methods for pesticides. Also, *Kontes
Quant Notes*, edited by J. Sherma is published quarterly by Kontes
of Vineland, New Jersey, and presents abstracts of recent papers
on TLC stressing in situ quantitation of components by color, fluorescence, and fluorescence quenching. This publication covers a variety
of topics and frequently includes abstracts on pesticide analysis.
Also, the application reviews published in *Analytical Chemistry*
are a very good source of information (8).

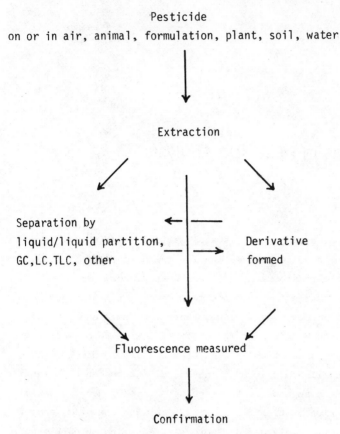

FIG. 10.1 Five schemes used in the analysis of pesticides by fluorescence. (Reprinted with permission from Ref. 7.)

Five fundamental approaches are generally used with fluorescence for determining pesticides in air, animals, formulations, plants, soil, and water. Figure 10.1 shows how the chemical is first extracted, after which the extract is subjected to one of five approaches before fluorescence measurement (7). The five approaches are (1) just additional dilution of the extract, (2) additional separation steps, (3) separation and preparation of a fluorescent derivative, (4) preparation of a fluorescent derivative, and (5) preparation of a fluorescent derivative followed by a separation from interfacing components (7).

OLDER METHODS

A chelate spray consisting of a 1:2 mixture of salicyl-2-aldehyde-
2-quinolylhydrazone (SAQH) and manganese(II) chloride has been used
for the determination of organothiophosphates by in situ fluorometry
on thin-layer chromatoplates. The SAQH-Mn reagent was more stable
to light and yielded more fluorescence than other chelate spray reagents studied earlier. The detection limits of several organothio-
phosphorous pesticides varied between 0.02 and 0.08 μg per spot
depending on the number and type of sulfur atoms. Parathion with
one sulfur atom had a detection limit of 0.08 μg per spot, and tri-
thion with three sulfur atoms had a detection limit of 0.02 μg per
spot. Reproducibility studies were carried out with 1-μg spots of
malathion and parathion, and the relative standard error was 2.4%
and 2.9%, respectively. The method was applied to water samples
spiked with guthion, and concentrations as low as 0.5 ppb could
be determined on a routine basis. The method offered a reasonable
alternative to gas-liquid chromatography (5). Later a method was
developed using the above approach for determining guthion in blue-
berries (5).

Flavones as spray reagents were an important addition to the
evaluation of pesticides (5,9,10). Flavones were chosen because
of their ability to be almost nonfluorescent in a nonpolar medium
and very fluorescent in a polar one. For example, if such compounds
were sprayed onto a surface of low polarity such as cellulose, the
background would be weak or nonfluorescent. The presence of a polar
compound on the surface would enhance the fluorescence of the spray
reagent and thus allow detection of a spot. It was found that 3-hy-
droxyflavones with the fifth position unsubstituted were good reagents. Fisetin was good in this respect. Fluorescence enhancement
of the flavone occurred with pesticides on the cellulose plates
(5).

Fisetin

Mallet and Frei (9) assessed the usefulness of this approach
to a wide range of pesticide systems such as s-triazines, organo-
phosphates, carbamates, chlorinated hydrocarbons, and others. They
also investigated herbicides and fungicides. The cellulose plates
with pesticides on them were sprayed with a 0.05% solution of fisetin
in isopropanol and the resulting fluorescence observed under long
wavelength ultraviolet radiation. The visual detection limits for
carbamates were from 0.06 to 0.01 μg. A number of organophosphorus
compounds were tested and gave detection limits from 0.04 to 0.02
μg. Fluorescence enhancement was also observed for s-triazines,
and visual detection limits were from 0.04 to 0.1 μg. The approach
was also suitable for chlorinated hydrocarbons but sensitivity was
less (∿0.1 μg) due to their lower polarity. Mallet and Frei (10)
reported a method for the detection and determination of organic
compounds by in situ fluorometry after separation on cellulose layers.
Proban, an organothiophosphorus pesticide, was used as a test sample
in the quantitative studies. The 3-hydroxyflavone, robinetin, was
found to give very good results and linear calibration curves between
0.1 and 10 μg were obtained. The utility of the method was tested
by determining proban in tap water at the 1-ppb level. Mallet and
Frei concluded that their approach would be useful for water samples
and checking pesticide formulations.

Belliveau and Frei (11) described the detection of a large number
of sulfur-containing compounds on silica gel chromatoplates. They
first exposed the chromatoplates to bromine vapor and then sprayed
the plates with 1,2-dichloro-4,5-dicyanobenzoquinone (DDQ). The
approach was based on two chemical reactions; first, the reaction
between sulfur in the pesticide and bromine with liberated acid and
second, the reaction of the acid with DDQ. Belliveau and Frei did
not measure detection limits, but the method was effective at the
1-μg level. Table 10.1 lists several compounds that were detected.
Because of the ultraviolet instability of DDQ a survey of other
pH sensitive fluorogenic spray reagents was made (5). DDQ yielded
the highest fluorescence with 8-hydroxyquinaldine giving almost the
same results. 8-Hydroxyquinoline, salicylaldoxine, and thiophene-2-
aldehyde-hydrazone were found to be useful agents.

TABLE 10.1 Organophosphorus Pesticides which Exhibit Blue Fluorescence on a Nonfluorescent Background after Exposure to Bromine for less than 1 Minute, and Spraying of DDQ

Pesticide	Structure[a]	Pesticide	Structure[a]
Baytex	$(MeO)_2-P(=S)-O-$ (phenyl with Me and SMe substituents)	Menazon	$(MeO)_2-P(=S)-S-CH_2-$ (triazine ring with two NH_2 groups)
Cygon	$(MeO)_2-P(=S)-S-CH_2-C(=O)-NH-Me$	Methyl trithion	$(MeO)_2-P(=S)-S-CH_2-S-$ (phenyl with Cl)
Diazinon	$(EtO)_2-P(=S)-O-$ (pyrimidine ring with Pr^i and Me substituents)	Papthion	$(MeO)_2-P(=S)-S-CH(phenyl)-C(=O)-OEt$
Dyfonate	$EtO-P(=S)(Et)-S-$ (phenyl)	Parathion	$(EtO)_2-P(=S)-O-$ (phenyl with NO_2)
Ethion	$(EtO)_2-P(=S)-S-CH_2-S-P(=S)-(OEt)_2$	Folithion	$(MeO)_2-P(=S)-O-$ (phenyl with Me and NO_2)

TABLE 10.1 (Continued)

Guthion	$(MeO)_2-P(=S)-S-CH_2-N=N-$ (benzene ring with C=O)	Proban	$(MeO)_2-P(=S)-O-$ (benzene ring) $-S(=O)(=O)-NH_2$
Imidan	$(MeO)_2-P(=S)-S-CH_2-N$ (phthalimide ring with two C=O)	Systox[b]	$(EtO)_2-P(=O)-S-CH_2-S-Et$
M-1703	$(MeO)_2-P(=S)-S-CH-C(=O)-O-CH_2-CH_2-F$ (with phenyl $-S(=O)-$)	Thimet	$(EtO)_2-P(=S)-S-CH_2-S-Et$
Malathion	$(MeO)_2-P(=S)-S-CH(C(=O)-OEt)-CH_2-C(=O)-OEt$	Trithion	$(EtO)_2-P(=S)-S-CH_2-S-$ (benzene ring) $-Cl$

[a] Me = $-CH_3$; Et = $-C_2H_5$; Pr^i = $-CH(CH_3)_2$.

[b] Needs exposure to bromine for a longer period of time.

Reprinted with permission from Ref. 11.

NEWER METHODS

In the methods described previously, the fluorescence was obtained
from reagents that were not very selective. These methods can be
somewhat inconvenient because the fluorescence is very similar for
all the compounds on a chromatoplate whether they are pesticides
or impurities. Newer approaches obtain the fluorescence directly
from the pesticides. In these cases, the material is naturally fluorescent, or it is converted into a species which fluoresces. The obvious advantage is that the fluorescence is particular to the compound
of interest. There are a limited number of pesticides that are
naturally fluorescent, but many can be converted to fluorescent
species by a variety of methods (3,5,6). Lawrence and Frei (3)
commented that most earlier fluorometric methods of pesticide analysis
used classical solution-fluorescence measurements. This approach
is not suitable for fluorescence analysis of pesticide residues
because of the rigorous cleanup required to minimize blank fluorescence. TLC work associated with fluorescence and fluorescence derivatization techniques has eliminated several problems of fluorescent
coextractives. TLC allows the means of separating the desired pesticide from impurities and gives more accurate results. After separation the pesticides can be determined by in situ fluorescence.

Mallet, Surette, and Brun (12) investigated the native fluorescence of several pesticides (Table 10.2). The fluorescence was measured directly from silica gel chromatoplates. They found that heating
the chromatoplate resulted in changing the excitation and emission
maxima to longer wavelengths. The visual limits of detection were
in the nanogram range for heated and unheated chromatoplates. The
fluorescence information obtained from heated and unheated chromatoplates enhances the selectivity of the approach, and in certain cases
fluorescence intensity was increased markedly. Brun, Surette, and
Mallet (13) expanded the above approach for the detection of eight
organophosphorus pesticides by in situ fluorometry on thin-layer
chromatoplates. They spotted the pesticides on silica gel chromatoplates and developed the plates with n-hexane/acetone (5:1). The

TABLE 10.2 List of Pesticides

Pesticide (Manufacturer)	Uses	Structure
Benomyl (Du Pont)	F	Methyl N-[1-(butyl-carbamoyl)-2-benzimidazole]carbamate
Coumatetralyl (Chemagro)	Ro	4-Hydroxy-3-(1,2,3,4-tetrahydro-1-naphthyl)-coumarin
Diphacinone (Velsicol)	Ro	2-Diphenylacetyl-1,3-indandione

Fuberidazole   F   2-(2^1-Furyl)-benzimidazole
(Chemagro)

Propyl isome   S   Dipropyl 5,6,7,8-tetrahydro-7-
(Penick)   methylnaphtho [2,3-d]-1,3-dioxole-
                           5,6-dicarboxylate

Quinomethionate   I,M,F   6-Methyl-2,3-quinoxalinediyl
(Chemagro)   cyclic S,S-dithiocarbonate

F = fumigant; I = insecticide; M = miticide; Ro = rodenticide; S = synergist; s = saturated.

Reprinted with permission from Ref. 12.

developed plates were heated at a variety of temperatures for several
periods of time. When the optimum temperature and time of heating
were known, fluorescence excitation and emission spectra were recorded.
Almost all the pesticides they investigated were nonfluorescent;
however, by heating the pesticide at about 200°C for a period of
time analytically useful signals were obtained. Detection limits
were in the nanogram to microgram range. The compounds studied
by them were azinphosmethyl, coumaphos, dursban, imidan, maretin,
menazon, phosalone, and zinophos.

Mallet and Surette (14) showed how selectivity could be increased
by using acid or base as spray reagents with heat treatment of chromatoplates. Pesticides investigated were coumatetralyl, diphacinone,
fuberidazole, methabenzthiazuron, naphalam, propyl isome, quinomethionate, rotenone, thioquinox, and warfarin. Generally, spraying
with a strong electrolyte, such as an acid or base, before heat
treatment did not increase the limit of detection substantially
compared to just heat treatment. However, sometimes a change in
spectra was noted with acids which improved the selectivity of the
approach, but almost always decreased the fluorescence intensity.
Alkali spray reagents proved most useful as fluorescence intensi-
fiers.

Volpe and Mallet (15) investigated the major fluorescent species
obtained on TLC chromatoplates after pyrolysis of coumaphos and
potasan samples on chromatoplates. They also studied the degradation of coumaphos in water. The major fluorescent products of coumaphos and potasan after heating on chromatoplates were identified
as chlorferone and 4-methyl-umbelliferone using TLC, infrared, mass,
and fluorescence spectral data. Fluorescence spectra were routinely
recorded from chromatoplates. In water, coumaphos degraded into
coroxon, the oxygen analog, and chlorferone, the hydrolysis product.
They also proposed an in situ TLC fluorescence method for the identification of potasan in technical coumaphos that is much simpler
than existing methods.

Caissie and Mallet (16) investigated aminocarb, captan, coumaphos, difolatan, diphacinone, diquat, fuberidazole, guthion, landrin,

amretin, mesural, MGK repellent II, mobam, morestan, paraquat, rotenone, salithion, and warfarin for their fluorogenic properties
on acidic and basic aluminum oxide layers. They compared fluorescence data before and after heat treatment of the pesticides on
chromatoplates. They viewed the fluorescent components under ultraviolet radiation, and fluorescence spectra were obtained directly
from the chromatoplates. All the pesticides except aminocarb showed
a detectable natural fluorescence on acidic aluminum oxide layers.
Heat treatment of the chromatoplates resulted in intensified fluorescence for most pesticides. Also, the combination of excitation and
emission wavelength maxima was characteristic for each pesticide,
making their approach very selective. Heat treatment on basic aluminum oxide improved substantially the fluorescence of the pesticides
in most instances. Interestingly, some pesticides yielded similar
fluorescence efficiency on both acidic and basic aluminum oxide
layers, others fluoresced only on one type of layer. Their results
showed that the fluorescence data could be used to characterize pesticides, and aluminum oxide offered an alternative to the use of silica
gel chromatoplates.

Surette and Mallet (17) employed inorganic salts for enhanced
fluorometric detection of pesticides by spraying inorganic salts
onto the chromatoplate or by incorporating the inorganic salts into
the chromatoplate. A typical salt was aluminum chloride. Detection
limits between 8 and 40 ng were achieved on silica gel H and silica
gel 60 layers. Hausler et al. (18) reported a quantitative in situ
fluorescence method for the plant growth regulator 3-indoleacetic
acid on silica gel chromatoplates. They sprayed the chromatoplates
with 5% methanolic dimethylsulfoxide to enhance fluorescence intensity.
The calibration curve was linear from 50 ng to 0.1 μg.

Lawrence and Frei (19) and Frei and Lawrence (20,21) investigated fluorogenic labeling of carbamates with dansyl chloride (5-di-
methylaminonaphthalene-1-sulfonyl chloride). They have published
an expanded treatment of this reagent with carbamates and other
pesticides (3). Generally, the procedure with carbamates involved

the hydrolysis of the pesticide. Hydrolysis products were labeled
with dansyl chloride and then detected and quantitated by TLC on
silica gel chromatoplates by in situ fluorometry. Lawrence and
Frei (19) reported the experimental conditions affecting the fluorescent dansyl derivatives of seven carbamates separated on silica
gel chromatoplates with chloroform. They sprayed the derivatives
with 20% triethanolamine in isopropanol and aqueous buffer sprays
to increase fluorescence intensity. Frei and Lawrence (20) used
dansyl chloride for labeling nine N-methyl carbamate insecticides
with a relative standard deviation of 4.5% at 5-500 ng per spot.
Calibration curves were linear up to 300-400 ng per spot. The pesticides were extracted from spiked water samples, derivatized, and
separated on silica gel or silica gel-Ag_2O layers with a benzene/ace-
tone (98:2) mobile phase prior to quantitation. Frei and Lawrence
(21) studied the labeling of the carbamate insecticides matacil
and zectran with dansyl chloride and NBD chloride (4-chloro-7-nitro-
benz[2,1,3]oxadiazole), followed by TLC and densitometry. The relative standard deviation for the methods was 3-5% for 15-300 ng per
spot. Silica gel chromatoplates were employed and dansyl derivatives were detectable at 1 ng per spot. Zectran was determined
at the 200 ppb level in a soil sample extract.

Frei-Hausler, Frei, and Hutzinger (22) reported conditions for
the fluorogenic labeling of chlorophenols with dansyl chloride.
Chlorophenols are used as wood preservatives and fungicides, and
the hydrolysis products of several pesticides and herbicides yield
compounds such as 2,4-dichlorophenol and 2,4,5-trichlorophenol.
The authors described the reaction conditions, chromatographic conditions, and instrumental conditions for in situ fluorescence measurements. Mass spectral information was also given for the identification of the dansyl derivatives.

Lawrence, Renault, and Frei (23) discussed the fluorogenic
labeling of organophosphate pesticides with dansyl chloride. The
resulting fluorescent derivatives were separated by high-performance
liquid chromatography (HPLC) and TLC. Fluorescence detection was

used in both cases. Silica gel G chromatoplates and a balanced-density slurry-packed silica gel (10 μm) column were employed in the separation experiments. They found for best reproducibility at least 50 ng per spot was necessary for in situ fluorometry. They found HPLC with fluorescence detection gave lower detection limits, and the time of analysis was much shorter than in situ fluorometry (Table 10.3). With in situ fluorometry, they exposed the chromatoplate to ultraviolet radiation for 60 min to enhance the fluorescence. This added substantial time to the in situ method, and is not normally done with other in situ methods. With other in situ methods, spotting, separation, and analysis times can be reduced so that the time of analysis would be similar to HPLC. Also, in other HPLC methods the retention times of the components can be substantially longer, increasing the analysis time of HPLC. Furthermore, with TLC several samples can be run simultaneously while with HPLC this cannot be done. More data are needed comparing the two approaches before detailed comparisons can be made.

Zakrevsky and Mallet (24) developed an in situ fluorometric method for the detection and quantitative analysis of fenitrothion, its breakdown products and other amine-generating compounds. The

TABLE 10.3 Comparison of Total Analysis Time for Eight Samples of Fenthion

	Time (min)	
Step	TLC	HPLC
Spotting (TLC)	20	--
Chromatography	60 (including drying and spraying)	64 (8 × 8 min)
Analysis	60 (UV exposure)	0 (done simultaneously with chromatography)
	30 (scanning)	
Total	170	64

Reprinted with permission from Ref. 23.

technique was based on the reaction of fluorescamine with primary
amines to yield highly fluorescent derivatives. Other pesticides
studied were aminotriazole, fenitrothion, parathion, and parathion-
methyl. Silica gel chromatoplates were employed in the separation
step, and a Turner III fluorometer with a Camag TLC scanner was
used for all quantitative fluorometric measurements. For the detection of compounds bearing a primary amino group, the chromatoplate
was first sprayed lightly with water/acetone (1:1) solution. Then,
the chromatoplate was sprayed with the fluorescamine reagent followed
by a second water/acetone spray until the plate was just moist.
Nitro compounds were reduced to their corresponding primary amines
by spraying the chromatoplate to saturation with a tin(II) chloride
solution, allowing the chromatoplate to stand for 5 min, and drying
in a stream of air. Acid on the plate was neutralized by spraying
with 2 M sodium carbonate solution. After the above treatment,
the chromatoplate was sprayed with fluorescamine. All quantitative
measurements were performed on the Turner fluorometer. Table 10.4
gives the fluorescence data of the fluorescamine derivatives studied.
It can be seen that the detection limits are in the nanogram range.
Also, in some cases, spraying the plate with a 10% solution of
ethanolamine solution enhanced the fluorescence. The reproducibility
of the approach was evaluated for aminofenitrothion and fenitrothion
using 0.4-μg spots. The average relative standard deviation was
9.2% for fenitrothion and 7.2% for aminofenitrothion. The authors
commented that the diffusion of spots and greater background irregularities on the chromatoplates, caused by spraying with the inorganic
reagents, explained the lower precision for fenitrothion compared
to aminofenitrothion. They concluded that the procedure described
was useful for the simultaneous detection and quantitation of breakdown products of fenitrothion in samples of natural water and the
procedure could be applied to several other pesticides. Also, provided that the pesticides to be measured were separated from coextractives, the in situ technique offered a simple and convenient
method of analysis.

TABLE 10.4   Fluorescence Properties of the Fluorescamine
Derivatives on Thin Layers of Silica Gel

Compound	Excitation maximum (nm)	Emission maximum (nm)	Limit of detection (µg)	Upper limit of linear range (µg)
Aminofenitrothion	382	500	0.01	0.5
4-Amino-3-methylphenol	382	492	0.02	0.5
Aminotriazole	382	488	0.01	0.5
Fenitrooxon	383	490	0.01	1.0
Fenitrothion	383	490	0.01	1.0
Parathion-methyl	390	510	0.01	1.0
3-Methyl-4-nitrophenol	385	490	0.08 (0.02[a])	1.0
Parathion	390	510	0.01	1.0

[a] Limit of detection after spraying with 10% triethanolamine in
ethanol.

Reprinted with permission from Ref. 24.

Young, Khan, and Marriage (25) described the fluorescence detection and determination of the herbicide glyphosate in plants
by its N-nitroso derivative using in situ fluorometry. Glyphosate
controls top growth and prevents shoot regrowth from the root or
rhizome system of widespread perennial weeds such as the Canada
thistle. The method involved: (1) aqueous extraction of glyphosate
from the sample, (2) preparation of the corresponding N-nitroso
derivative with sodium nitrite at pH 3, (3) TLC separation, (4) photo-
lytic cleavage of the nitroso derivative with ultraviolet radiation
to produce a primary amine, (5) formation of a fluorophor with fluorescamine, and (6) in situ measurement of fluorescence. The advantages
of the method were: (1) sensitivity, as little as 10 ng could be
detected; (2) selectivity, interferences were only from surviving
primary amines or compounds yielding primary amines after nitrosation
and ultraviolet irradiation; (3) simplicity, extensive cleanup steps

were not required; (4) speed, once the initial extraction was completed, the remaining steps were done in several hours and many
samples could be analyzed simultaneously. The authors also mentioned
that other reported methods are more complex and time-consuming.

Sherma and Touchstone (26) developed methods for determining
chloramben herbicide. Fluorescamine was used to produce fluorescent
spots on silica gel chromatoplates or alternatively N-(1-naphthy)ethy-
lene diamine dihydrochloride was used to give nonfluorescent red spots
on the chromatoplates. A Schoeffel SD 3000 spectrodensitometer was
employed in the fluorescence-reflection mode for obtaining the fluorescence data or in the double-beam transmission mode for absorption
measurements. They carried out recovery studies with tomato and
lima bean samples obtained from a supermarket. The samples were
fortified with chloramben at 0.1-ppm levels after no pesticide was
found in blank runs. Chloramben was isolated from tomatoes and
lima beans by methods published previously with slight modifications.
Concentrated extracts were separated on silica gel chromatoplates.
The authors found the resulting calibration curve for chloramben
sprayed with fluorescamine reagent to be linear from 20 to 500 ng,
and the lowest detectable amount was 20 ng. Excellent reproducibility
was obtained for a given amount of pesticide. Interestingly, the
chromogenic method was found to be one of the most sensitive ever
encountered for detection and quantitation of colored TLC spots.
The linearity of the calibration curve was from 4 to 80 ng, and the
detection limit was 4 ng. Both the fluorogenic and chromogenic methods
were simple, sensitive, and rapid. The sensitivity was at least
as good as microcoulometric gas chromatographic methods, and both
reagents were selective for compounds with $-NH_2$ groups. The TLC
separation step was rapid (30 min). The researchers emphasized
that quantitative TLC appears to have great potential as an alternative to gas chromatography and HPLC in the determinative step
in pesticide analysis.

MacNeil and Hikichi (27) developed a rapid method for the determination of benomyl (methyl-1-(butylcarbamoyl)benzimidazole-2-
ylcarbamate) on cherries by silica gel TLC and in situ fluorometry.

Linear calibration curves were from 50 to 500 ng per spot and the
detection limit was about 20 ng. The method was applied to commercially treated cherries, and it was sensitive to less than 1 ppm
benomyl by weight on the sample surface. Reproducibility studies
showed the analyses could be performed with a relative standard
deviation of about ±10% between 200 and 800 ng. Several analyses
could be done simultaneously, and the combination of TLC and in situ
fluorometry provided a simple and rapid method for the analysis
of benomyl residues on cherries.

Sherma (28) reported on the quantitation of benomyl and its
metabolites by thin-layer densitometry. Both fluorescence and fluorescence quenching on silica gel chromatoplates were studied. In contrast to the work by MacNeil and Hikichi (27) discussed previously,
Sherma found very weak fluorescence signals from benomyl adsorbed
on silica gel chromatoplates. MacNeil and Hikichi used different
commercial instrumentation and commercial chromatoplates compared
to Sherma's approach. Sherma concluded it was better to employ
fluorescence quenching for benomyl. He used commercial chromatoplates that contained a fluorescent indicator. Sherma obtained
quenching data on benomyl, the principal degradation product of
benomyl, MBC(methyl-2-benzimidazolecarbamate), and a minor degradation product in plants, 2-AB(2-aminobenzimidazole). Both MBC and
2-AB gave weak native fluorescence on silica gel chromatoplates.
The fluorescence quenching calibration curves were linear over a
wide range of concentrations. For example, a MBC calibration curve
had a range form of 0-6 µg. The limits of detection for benomyl,
MBC, and 2-AB were 600 ng, 200 ng, and 500 ng, respectively. Sherma
stated that quantitative TLC determinations are selective, sensitive, rapid, and simple, and gave an attractive alternative to read-
out by gas chromatography, HPLC, or spectrophotometry for residue
analysis.

REAL-LIFE SAMPLES

There have been several applications of in situ fluorometry to "real-
life" samples and some of these have already been discussed. Below
several selected examples are discussed.

MacNeil, MacLellan, and Frei (29) determined dimethoate and
malathion in lettuce. The phosphate insecticides were extracted
from lettuce. The resulting concentrate was spotted directly on
silica gel chromatoplates for TLC of dimethoate, but a Florisil
column cleanup step was needed to remove fluorescent pigments before
quantitation of malathion. The separated insecticides on chromatoplates were sprayed with a palladium chloride-calcein reagent and
stored in a closed tank 18-24 hr to develop full fluorescence.
A 10-100 ng per spot linear calibration curve was obtained for dimethoate. Dimethoate-spiked extracts averaged recoveries from 85-105%.

Zakrevsky and Mallet (30) developed an in situ fluorometric method
for determination of bayrusil in foodstuffs. They expressed the
need for a method that was more simple than existing gas chromatographic methods. Bayrusil is used for the control of biting and
sucking insects in field crops and fruits. Samples of apples, lettuce, tomatoes, cucumbers, and beans were investigated. They were
extracted first, then a Florisil column cleanup step was employed,
and finally two-dimensional TLC was carried out. Before the second
development in the TLC separation, the chromatoplate was heated at
100°C for 30 min. This step induced the fluorescence of the pesticide and allowed sensitive detection and quantitation of bayrusil.
The fluorescence was very stable, and the limit of detection of
bayrusil was about 6 ng per spot. Recoveries of bayrusil at the
0.02 ppm level were very good and ranged from 96-107% recovery.

Zakrevsky and Mallet (31) proposed in situ fluorometry as an
alternative to gas chromatography for determining coumaphos (0,0-
diethyl-0-(3-chloro-4-methyl-2-oxo-2H-1-benzopyran-7-yl)phosphoro-
tioate) and its oxygen analog, coroxon, in eggs. With the gas chromatography approach, interfering substances from various sources often
made accurate quantitative determination difficult, and additional
separation steps were needed. The method developed by Zakrevsky
and Mallet allowed for the simultaneous determination of coumaphos
and coroxon in eggs in the ppb range. Initially the sample was
extracted and then the extract was purified by liquid-liquid partition

followed by Florisil column chromatography. Finally, two-dimensional
TLC on silica gel was carried out. The developed chromatoplate
was heated at 200°C for 20 min to increase the fluorescence. Coumaphos
and coroxon had identical fluorescence characteristics, but they
had widely different R_f values on silica gel, which aided in identification of the components. Percentage recovery values compared
well with gas chromatography results at 0.10 and 0.02-ppm levels and
were satisfactory for the types of samples they investigated. The
limit of detection for both compounds was 0.001 µg per spot, and
calibration curves were linear to 10 µg per spot. Both gas chromatography and in situ fluorometry took about the same time, but fewer
manipulations were required by TLC and the cost was less.

Thiabendazole [2-(4-thiazolyl)benzimidazole (TBA)] is used
as a fungicide in banana and citrus cultivation. Otteneder and Hezel
(32) developed a routine quantitative method for TBA by fluorometric
evaluation of thin-layer chromatograms. Fruit samples were initially
extracted, and the concentrated extract was spotted on silica gel
G chromatoplates and developed with ethyl acetate/methyl ethyl ke-
tone/formic acid/water (5:3:1:1). The developed chromatoplates were
then evaluated fluorometrically with a Zeiss chromatogram spectrophotometer. Otteneder and Hezel studied the photodecomposition
of TBA and the influence of rind substances on fluorescence. Recovery of TBA added to oranges ranged from 93 to 100%, and for grapefruit from 91 to 103%. The linear range for TBA calibration curve
was typically from 50 to 250 ng, and for 12 fruit samples investigated the TBA content ranged from 0.69 to 3.60 ppm. The authors
stated that in situ evaluation was faster and required less substance
per spot than elution and subsequent photometry.

Francoeur and Mallet (33) described a simple method for the
determination of quinomethionate (6-methylquinoline-2,3-diyldithio-
carbonate) in crops by in situ fluorometry. Quinomethionate is
used for control of mildew, mites, and mite eggs on crops such as
apples and pears. The pesticide residue was extracted with acetonitrile and partitioned in petroleum ether. After separation of coextractives by TLC, the fluorescence of quinomethionate was measured

directly from a silica gel chromatoplate. An average of 89% recovery
was obtained at the 0.05-ppm level in apples, peaches, pears, and
tomatoes. The procedure developed did not necessitate preliminary
cleanup, and most of the interfering materials were eliminated by
one-dimensional TLC. The authors concluded that the approach was
rapid and could be used as a new method for the determination of
quinomethionate in crops or as a confirmation technique for gas
chromatography results.

Francoeur and Mallet (34) described a simple method for the
simultaneous quantitative determination of the fungicides captan
and captafol in apples and potatoes. The fungicides were extracted
with acetonitrile and then partitioned into a solution of methylene
chloride in petroleum ether. They were separated from each other,
and coextractives with silica gel chromatoplates were impregnated
with aluminum chloride. The developed chromatoplates were sprayed
with 0.1 M sodium chlorate and heated. The aluminum chloride acted
as a catalyst to help fragmentize the fungicide molecules during
heating and the sodium chlorate acted as a reactant or complexing
agent to produce the fluorescent species. The minimum detectable
amount was 0.02 ppm, and the relative standard deviation was 7-8%
at 0.2 ppm and 10-15% at 0.05 ppm. At the 0.2-ppm level in test
crops the percentage recovery was greater than 90%.

GASEOUS ELECTRICAL DISCHARGE

Davies and Pretorius (35) investigated the fluorescence of pesticides on silica gel HPTLC chromatoplates by use of a gaseous electrical discharge technique. The apparatus used was described in
Chapter 3 and was developed by Shanfield, Hsu, and Martin (36).
Davies and Pretorius showed that the optimum conditions for the
electrical discharge experiment were 4 min in the electrical discharge chamber; 8 min in a forced draught oven at 130°; and 12 min
under 350-nm ultraviolet radiation. Table 10.5 gives detection

TABLE 10.5 Detection Limits of Pesticides on HPTLC Silica Gel Plates

Pesticide	UV only				Heat plus UV				ED plus N_2 plus heat plus UV			
	1	0.5	0.2	0.05	1	0.5	0.2	0.05	1	0.5	0.2	0.05
Diuron	+	+	-	-	+	+	-	-	+	+	+	-
Atrazine	+	+	+	-	+	+	+	-	+	+	-	-
Picloram	+	+	-	-	+	+	-	-	+	+	-	-
γ-BHC	-	-	-	-	-	-	-	-	-	-	-	-
Endosulfan	-	-	-	-	-	-	-	-	-	-	-	-
p,p'-DDT	-	-	-	-	-	-	-	-	+	+	-	-
PCP	-	-	-	-	-	-	-	-	+	+	-	-
Monocroto-phos	+	+	-	-	+	+	-	-	+	+	-	-
Dimethoate	+	+	-	-	+	+	-	-	+	+	-	-
Fenitro-thion	+	+	-	-	+	+	-	-	+	+	+	-
Diazinon	+	+	-	-	+	+	-	-	+	+	-	-
Propoxur	-	-	-	-	-	-	-	-	+	+	-	-
Methyl Gusathion	+	+	+	+	+	+	+	+	+	+	+	+
Perthane	-	-	-	-	-	-	-	-	+	+	-	-

ED = Electrical discharge; UV = ultraviolet radiation (350 nm);
1, 0.5, 0.2, and 0.05 = amount spotted in micrograms; + = detectable visually; - = not detected.

Reprinted with permission from Ref. 35.

limits of several pesticides on HPTLC silica gel chromatoplates.
Davies and Pretorius stated that detection limits by other techniques
for pesticides were lower. However, the attractiveness of the electrical discharge technique lies in its simplicity and application
to a wide range of pesticides. Also, the future development of
high-performance silica gel and alumina-coated plates without binder
should improve the detection limits.

CONCLUSIONS

In their review on in situ fluorescence spectroscopy of pesticides
and other pollutants, Mallet, Belliveau, and Frei (5) emphasized
how the in situ approach has proven to be a powerful tool for the
qualitative and quantitative determination of minute quantities of
organic compounds. They compared analytical methods for organothiophosphorus pesticide residues. Enzymatic methods (37,38) are, in
some cases, more sensitive than gas chromatographic methods. The
fluorogenic methods described by Mallet, Belliveau, and Frei lie
somewhere between enzymatic and chromogenic detection methods in
terms of sensitivity. The main justification for the application
of fluorogenic reagents for detection of pesticides on thin-layer
chromatograms is that the fluorescence can be quantitated, which
is not necessarily the case with enzymatic reactions on TLC. Reflectance spectroscopy has been used for in situ quantitation of thio-
phosphates, but the sensitivity is at least 10 times less sensitive
than many other methods for pesticides. Quantitative TLC methods
that involve scraping the spot off the plate and extracting the adsorbent with a suitable solvent result in increased time, possible
losses during scraping, and possible incomplete extraction compared
to in situ methods. Solution fluorometric methods can be very sensitive in some cases but can be affected substantially by the presence
of extraneous impurities which necessitates preliminary cleanup steps.
Many times extraneous impurities are not a serious problem with
TLC methods. Mallet, Belliveau, and Frei pointed out that generally
the precision and accuracy of in situ methods are comparable to
gas chromatographic methods. Also, multiple spotting is possible
with TLC. This is in contrast to gas chromatography in which samples
are injected sequentially. With gas chromatography, preliminary cleanup
steps are often necessary, but TLC itself is a useful cleanup technique. For TLC this results in time saving. Another advantage of
in situ techniques is the relatively low cost of equipment and operation (5).

Lawrence and Frei (39) have compared quantitative HPLC and quantitative TLC. They commented that usually the chromatography and

detection processes can be better controlled in HPLC. HPLC is usually preferred for quantitation; however, TLC, because of its simplicity, flexibility, and cheapness, has its place as a quantitative
chromatographic method. Lawrence and Frei did not comment on HPTLC,
and certainly with future developments in HPTLC quantitative HPTLC
should become equivalent to quantitative HPLC in many situations.
Generally presample cleanup is not as stringent for TLC, while HPLC
requires reasonably clean samples for column injection (40). Lawrence
and Frei (39) discussed fluorometric derivatization of pesticides
and stated many derivatization techniques could be applied with
both quantitative TLC and HPLC. It appears much additional
developmental work can be done for the analysis of pesticides with
both TLC and HPLC.

REFERENCES

1. J. J. Aaron, E. Kaleel, and J. D. Winefordner, *J. Agr. Food
    Chem.*, in press.

2. J. J. Aaron and J. D. Winefordner, *Analysis*, 7, 168 (1979).

3. J. F. Lawrence and R. W. Frei, *J. Chromatogr.*, *98*, 253 (1974).

4. R. W. Frei and J. Lawrence, *Methods in Residue Analysis*,
    Vol. IV, A. S. Tahori, ed., Gordon and Breach Science Publishers,
    New York, 1971, pp. 193-202.

5. V. N. Mallet, P. E. Belliveau, and R. W. Frei, *Residue Re-
    views*, Vol. 59, F. A. Gunther and J. D. Gunther, eds., Springer-
    Verlag, New York, 1975, pp. 51-90.

6. J. D. MacNeil and R. W. Frei, *J. Chromatogr. Sci.*, *13*, 279 (1975).

7. R. J. Argauer, *Analytical Methods for Pesticides and Plant
    Growth Regulators*, Vol. IX, G. Zweig and J. Sherma, eds., Aca-
    demic Press, New York, 1977, pp. 101-136.

8. W. Thornburg, *Anal. Chem.*, *51*, 196R (1979).

9. V. N. Mallet and R. W. Frei, *J. Chromatogr.*, *56*, 69 (1971).

10. V. N. Mallet and R. W. Frei, *J. Chromatogr.*, *60*, 213 (1971).

11. P. E. Belliveau and R. W. Frei, *Chromatographia*, *4*, 189 (1971).

12. V. N. Mallet, D. P. Surette, and G. L. Brun, *J. Chromatogr.*,
    79, 217 (1973).

13. G. L. Brun, D. P. Surette, and V. N. Mallet, *Int. J. Environ.
    Anal. Chem.*, *3*, 61 (1973).

14. V. N. Mallet and D. P. Surette, *J. Chromatogr.*, *95*, 243 (1974).

15. Y. Volpe and V. N. Mallet, *Anal. Chim. Acta, 81*, 111 (1976).

16. G. E. Caissie and V. N. Mallet, *J. Chromatogr.*, *117*, 129 (1976).

17. D. P. Surette and V. N. Mallet, *J. Chromatogr.*, *107*, 141 (1975).

18. M. Hausler, J. D. MacNeil, R. W. Frei, and O. Hutzinger, *Mikrochim. Acta*, 43 (1973).

19. J. F. Lawrence and R. W. Frei, *J. Chromatogr.*, *66*, 93 (1972).

20. R. W. Frei and J. F. Lawrence, *J. Chromatogr.*, *67*, 87 (1972).

21. R. W. Frei and J. F. Lawrence, *J. Assoc. Offic. Anal. Chem.*, *55*, 1259 (1972).

22. M. Frei-Hausler, R. W. Frei, and O. Hutzinger, *J. Chromatogr.*, *84*, 214 (1973).

23. J. F. Lawrence, C. Renault, and R. W. Frei, *J. Chromatogr.*, *121*, 343 (1976).

24. J. G. Zakrevsky and V. N. Mallet, *J. Chromatogr.*, *132*, 315 (1977).

25. J. C. Young, S. U. Khan, and P. B. Marriage, *J. Agric. Food Chem.*, *25*, 918 (1977).

26. J. Sherma and J. C. Touchstone, *Chromatographia, 8*, 261 (1975).

27. J. D. MacNeil and M. Hikichi, *J. Chromatogr.*, *101*, 33 (1974).

28. J. Sherma, *J. Chromatogr.*, *104*, 476 (1975).

29. J. D. MacNeil, B. L. MacLellan, and R. W. Frei, *J. Assoc. Offic. Anal. Chem.*, *57*, 165 (1974).

30. J. G. Zakrevsky and V. N. Mallet, *Bull. Environ. Contam. Toxicol.*, *13*, 633 (1975).

31. J. G. Zakrevsky and V. N. Mallet, *J. Assoc. Offic. Anal. Chem.*, *58*, 554 (1974).

32. H. Otteneder and V. Hezel, *J. Chromatogr.*, *109*, 181 (1975).

33. Y. Francoeur and V. N. Mallet, *J. Assoc. Offic. Anal. Chem.*, *59*, 172 (1976).

34. Y. Francoeur and V. N. Mallet, *J. Assoc. Offic. Anal. Chem.*, *60*, 1328 (1977).

35. R. D. Davies and V. Pretorius, *J. Chromatogr.*, *155*, 229 (1978).

36. H. Shanfield, F. Hsu, and A. J. P. Martin, *J. Chromatogr.*, *126*, 457 (1976).

37. C. E. Mendoza, P. J. Wales, H. A. McLeod, and W. P. McKinley, *Analyst, 93*, 34 (1968).

38. A. M. Gardner, *J. Assoc. Offic. Anal. Chem.*, *54*, 517 (1971).

39. J. F. Lawrence and R. W. Frei, *Chemical Derivatization in Liquid Chromatography*, Vol. 7, Elsevier Scientific Publishing Co., New York, 1976, pp. 37-38, 186-198.

40. H. J. Issaq and E. W. Barr, *Anal. Chem.*, *49*, 83A (1977).

11

Applications in Food Analysis and Pharmaceutical Process Control

Applications in food analysis and pharmaceutical analysis will be considered in this chapter. Solid surface fluorescence analysis has been applied extensively to the analysis of *mycotoxins* (toxic metabolites of molds) in foods. Several other applications to food analysis have appeared in the literature and some of these applications will also be discussed. The applications of solid surface luminescence in the pharmaceutical industry will be considered from the point of view of process control and the analysis of pharmaceutical products. The analysis of drugs of abuse, and drugs and pharmaceuticals in human body fluids will be discussed in Chapter 12. The reader is referred to the application reviews published in *Analytical Chemistry* for a general review on food analysis and pharmaceutical analysis (1,2).

CITRUS OILS AND JUICES

Madsen and Latz (3) investigated qualitative and quantitative in situ fluorometry of citrus-oil components separated on thin-layer chromatoplates. Expressed citrus oils contain crystalline components that can be separated by thin-layer chromatography (TLC) and exhibit fluorescence under ultraviolet radiation. Latz and Madsen (4) reported an extensive examination of the luminescence properties of

expressed lime oil and coumarin derivatives isolated from expressed
lime oil. They showed 5-geranoxy-7-methoxycoumarin and 5,7-dimeth-
oxycoumarin were responsible for the fluorescence characteristics
of lime oil (4). Madsen and Latz used both fluorescence and fluorescence quenching profiles from thin-layer chromatoplates to characterize citrus oils. An Aminco-Bowman spectrophotofluorometer and
an American Instrument Co. motorized thin-film scanner were employed.
They presented fluorescence emission and fluorescence quenching
profiles of components distributed on chromatoplates for samples
of expressed citrus oils (Bergamont, California grapefruit,
Persian lime, and West Indian lime). Also, fluorescence emission
and fluorescence quenching profiles were given for expressed orange
oils (West Indian bitter orange, North American standard orange,
and Florida orange). Profiles were also presented for California
lemon oil, California grapefruit-oil mixture, Florida lime oil,
and California lemon oil. Detailed comparisons were made among the
profiles to characterize the samples. For example, the citrus oils
exhibited profiles which were generally characteristic of each particular kind of oil. Also, comparison of the emission and quenching
profiles of the citrus oils indicated an increased complexity of
expressed lemon and lime oils compared to other oils. Lemon and
lime oils exhibited similar profiles but showed variations which
allowed differentiation between the two. Many other comparisons
were made. Madsen and Latz made tentative identification of the
crystalline components responsible for many of the peaks in the emission and quenching profiles of Indian lime oil and expressed California
lemon oil. Identifications were based on results obtained from
in situ fluorometry and visualization of components under ultraviolet
radiation. Some of the compounds identified were 5-geranoxypsoralen,
5,7-dimethoxycoumarin, and 5,8-dimethoxypsoralen. Madsen and Latz
investigated the experimental parameters for direct quantitative
fluorometry of citrus-oil components on chromatoplates. Some of
the parameters studied were the influence of time on fluorescence
intensity, component spot alignment, and relationship of profile

peak area and peak height to concentration for obtaining linear analytical curves. They also determined 5-geranoxy-7-methoxycoumarin and 5,7-dimethoxycoumarin in citrus oils by in situ fluorometry. Their work illustrates the wealth of information that can be obtained from complex samples by solid surface fluorescence analysis.

Tatum and Berry (5) developed a visual comparison method for estimating the limonin content in citrus juices. Limonin contributes to the bitterness of grapefruit and orange juice. They were interested in a quick, simple method for the estimation of limonin concentration in citrus juice for quality assurance programs. Sixteen mobile phases were developed for the separation of limonin from citrus juices. The mobile phases were used with silica gel chromatoplates in various combinations to separate limonin from a variety of samples. Tatum and Berry employed two spray reagents: (1) 10% sulfuric acid in ethanol, and (2) 2% sulfuric acid, 1% p-dimethyl-aminobenzaldehyde in ethanol. After separation of limonin, the chromatoplates were sprayed with (1) or (2) until lightly wet and placed in an oven at 125°C for 6 min. After the chromatoplate was cooled, the concentration was estimated by comparing either visible color or fluorescence with limonin standards. The fluorescence of limonin after spraying appeared red to brown under ultraviolet radiation. The accuracy and precision of the method were good, and the limit of detection was in the parts per million range. The authors used six judges to evaluate the limonin content in samples to show that most technically trained people could determine limonin by the method and thus show its feasibility as a potential industrial quality-control method. They concluded most people could use their method successfully. The method was applicable to grapefruit and orange juices, fresh juice, commercial single-strength canned juice and orange juice concentrates.

CHLOROGENIC ACID, FLAVONOL GLYCOSIDES, HISTAMINE, OAT CULTIVARS, AND GIBBERELLINS

Van Buren, de Vos, and Pilnik (6) developed methods for the in situ fluorometric determination of the phenolic compounds, chlorogenic

acid and flavonol glycosides in apple juice. There has been considerable interest in the phenolic compounds of apples because these
compounds influence the color and clarity of the juice. The authors
adapted methods developed by Ibrahim (7) for the measurement of
chlorogenic acid and flavonol glycosides in apple juice. Ibrahim
investigated the quantitative in situ fluorometry of plant phenolic
compounds on chromatoplates. Van Buren, de Vos, and Pilnik employed
cellulose chromatoplates and a Vitatron TLD 100 flying spot densitometer in their work. The percent recovery of chlorogenic acid
in apple juices was between 91 and 102%, and the standard error
of the mean ranged from ±4 to ±8%. The method for flavonol glycosides in apple juice gave a range of standard error of the mean
of ±0.03 to ±0.18%. The authors commented on the advantages of
their method compared to other methods; separations were easier,
losses of materials were minimized, and a rough visual estimate
of concentrations could be made before densitometric readings.

There is considerable interest in histamine as a possible cause
of food poisoning. Lieber and Taylor (8) compared thin-layer chromatographic detection methods for histamine from food extracts. They
investigated the specificity and sensitivity of four reagents, ninhydrin, o-phthalaldehyde (OPA), fluorescamine, and o-diacetylbenzene
(DAB) for detection of histamine on thin-layer chromatoplates after
separation with several mobile phases. The ninhydrin reagent was
the most sensitive, with a detection limit of 0.4 nmoles of histamine.
The DAB and fluorescamine reagents had the most potential as a hista-
mine-specific visualization method. OPA was unsatisfactory as a
reagent because the fluorescent spots were unstable. Lieber and
Taylor tested 19 biogenic amines other than histamine. Ninhydrin
reacted with 17, fluorescamine with 8, and DAB with 15. As mentioned,
OPA was too unstable as a spray reagent. From previous work and
the results mentioned above, Lieber and Taylor (8,9) concluded TLC
provided a rapid means for the determination of potentially toxic
levels of histamine in foods. Also, TLC combined with the visualization reagents will simplify and shorten the analysis time for histamine determinations in quality-control laboratories and regulatory
laboratories.

Brach and Baum (10) investigated oat cultivars by fluorescence spectroscopy. Cultivars within a crop possess qualitative and quantitative differences that determine market value. Plant breeders and farmers need reliable identification means to ensure that the variety of crop grown is the variety intended. Brach and Baum constructed a fluorescence spectrophotometer and employed it to measure fluorescence reflected from samples of oats that were dehulled and ground and placed in a 10-mm path length cuvette. The samples were excited with wavelengths from 200 to 290 nm. Most of the fluorescence emission spectra were recorded from 300 to 500 nm. The results showed that there were amplitude differences in the fluorescence spectra for different varieties of samples. The authors were able to discriminate between cultivars by four peak points and four valley points on the spectra. Experiments were done with several hundred samples, and the peaks and valleys always occurred at the respective wavelengths. The authors speculated some of the fluorescence was due to folic acid, pyridoxine, riboflavin, and thiamine in the oat samples. Six varieties of oats were used to test the characterization procedure, and several hundred samples were tested. Brach and Baum used a statistical method based on the peak and valley data points from the fluorescence spectra to successfully discriminate the oat cultivars from one another.

Winkler (11) has presented methods for the determination of gibberellins in apples. Gibberellins, as naturally occurring plant growth regulators, affect cell elongation and promote cell division. Gibberellins A_4 and A_7 are naturally occurring in apple seeds and are also present in mature apple pulp. Gibberellins A_4A_7 were extracted from apple pulp with pH 7 buffered aqueous acetone. After the extract was concentrated, the solution was acidified, and A_4A_7 reextracted into ethyl acetate. The dried residue was subjected to preparative TLC, and the A_4A_7 zone was eluted, concentrated, and chromatographed in a TLC system that separates A_4 from A_7. The final chromatoplate was sprayed with sulfuric acid solution to develop fluorescence of A_4 and A_7. The A_4 and A_7 were then quantitated by in situ fluorometry. The procedure outlined above can quantitate A_4 and A_7 residues in apple pulp at levels as low as 1 ppb. Percent recoveries at the pbb level were as follows: A_4A_7, 59–89%; A_4, 50–76%; and A_7, 67–93%. The previous percent recovery values are very good, considering the small amount of gibberellins and the complexity of the samples.

N-NITROSAMINES AND N-NITROSAMINO ACIDS

Young (12) described a procedure for the detection and determination of N-nitrosamines (NAs) on thin-layer chromatoplates. Many NAs are known to be potent carcinogens. The NAs on activated silica gel or aluminum oxide chromatoplates were irradiated with ultraviolet radiation. The irradiation step yielded primary and secondary amines. After irradiation, the chromatoplates were sprayed with fluorescamine reagent which gave both fluorescent and nonfluorescent products. In situ fluorometric determinations were made with a Hitachi Perkin-Elmer Model MFP-2A fluorescence spectrophotometer equipped with a TLC accessory. A 15-W General Electric 18-in. germicidal lamp was used to irradiate the samples. The visual detection limits and R_f values are given in Table 11.1 for the several N-nitrosamines studied. Calibration curves were obtained for several NAs directly from chromatoplates, and nearly linear calibration curves were obtained on a log-log scale.

In later work, Young (13) reported on the detection and determination of N-nitrosamino acids by TLC using fluorescamine. The method was applied to samples of bacon. N-nitrosamino acids (NAAs) are thought to be potential precursors to the carcinogenic N-nitrosamines. Standards of NAAs were separated by TLC and irradiated with ultraviolet radiation as described above (12). The chromatoplates were then sprayed first with fluorescamine solution and then with triethanolamine solution. Fluorescence intensities were measured from the chromatoplate. Bacon samples were submitted to a procedure published previously. The bacon samples were ground, extracted, and

TABLE 11.1 Thin-Layer Chromatographic R_f Values and Visual Fluorescence Detection Limits of N-Nitrosamines on Silica Gel and Aluminum Oxide

		R_f value				Detection limit	
		Silica gel			Aluminum oxide		
No.	N-Nitrosamine	A	B	C	D	ng	pmoles
1	Dimethyl	0.35	0.44	0.24	0.44	500[a]	7000
2	Diethyl	0.44	0.60	0.39	0.52	10-40[b]	100-400
3	Dipropyl	0.58	0.65	0.49	0.58	7-10	50-75
4	Dibutyl	0.67	0.67	0.53	0.58	9-12	60-90
5	Dipentyl	0.73	0.68	0.56	0.66	4-6	20-30
6	Dihexyl	0.76	0.76	0.57	0.68	9-12	40-55
7	Diheptyl	0.77	0.78	0.58	0.69	9-12	35-50
8	Dioctyl	0.79	0.79	0.59	0.70	8-11	30-40
9	Diallyl	0.62	0.68	0.50	0.55	10-12	90-100
10	Di-iso-butyl	0.69	0.69	0.56	0.58	8-11	50-75
11	Dicyclohexyl	0.66	0.76	0.52	0.59	12-14	55-70
12	Dibenzyl	0.69	0.77	0.53	0.61	15-18	65-80
13	Methyl, butyl	0.45	0.58	0.38	0.54	7-10	65-85
14	Ethyl, butyl	0.55	0.62	0.47	0.55	8-11	60-90
15	Methyl, phenyl	0.63	0.67	0.49	0.59	17-20	125-150
16	Propyl, phenyl	0.70	0.73	0.56	0.63	10-12	60-75
17	Ethyl, benzyl	0.56	0.65	0.58	0.58	16-20	100-125
18	Phenyl, benzyl	0.70	0.75	0.55	0.60	7-10	35-50
19	Pyrrolidinyl	0.41	0.50	0.36	0.51	20-40[b]	200-400
20	Piperidinyl	0.22	0.34	0.20	0.39	17-20	150-175
21	Diphenyl	0.71	0.75	0.57	0.63	nf	nf
22	Morpholinyl	0.21	0.33	0.23	0.39	nf	nf
23	N-Methyl piperazinyl	0.03	0.03	0.06	0.28	nf	nf
24	Carbazolyl	0.76	0.79	0.60	0.60	nf	nf

Solvent systems A, B, and C, hexane/ether/methylene chloride (4:3:2), (5:7:10), and (10:3:2), respectively; system D, ethyl acetate/hexane (4:1). Detection limit is the minimum amount of an N-nitrosamine

TABLE 11.1   (Continued)

(NA) that ultimately gives a detectable visual fluorescence.   Range
due to difference in visual acuity of the two observers.   Determined
on activated silica gel by spotting NA, developing, irradiating, with
UV light, spraying with fluorescamine reagent, and viewing under
longwave UV light.   nf = Nonfluorescent products.

[a] If N-nitrosodimethylamine was not developed, then detection limit
was 10 ng (150 pmoles).   Detection limits improved only slightly
(ca. 25%) when nonvolatile NAs were not developed.

[b] Range due to relatively high volatility of NA.

Reprinted with permission from Ref. 12.

then filtered.   The evaporated filtrate was purified by column chromatography and then analyzed by in situ TLC.   Detection limits and
R_f values for the three NAAs studied are given in Table 11.2.   Two-
dimensional TLC and fluorescence detection were used to confirm
the identity of NAAs in bacon samples.   Two brands of uncooked bacon
were examined for NAAs.   One sample of one of the brands showed
N-nitrosoproline.   Young employed infrared spectroscopy and gas
chromatography/mass spectroscopy to examine extracts from bacon
samples and emphasized how the TLC fluorescence approach complemented
these methods.   Also, conformers of some NAAs were separated by
TLC and their ratio determined directly from the chromatoplate.
Agreement between Young's method and a published NMR method was
good.

Cross, Bharucha, and Telling (14) developed methods for the
identification and determination of volatile N-nitrosamines in cooked
bacon fat.   They emphasized that inexpensive screening methods are
needed for N-nitrosamines in foodstuffs.   In their approach, separation steps were based on double distillation and a cleanup step on
a silica gel column.   Two different methods were employed in the
determination of N-nitrosamines, a colorimetric method and an in
situ fluorometric method.   In the colorimetric method, nitrite was
released from N-nitrosamines with hydrobromic acid in acetic acid
and then reaction of nitrite was released with N-1-naphthylethyl-
enediamine dihydrochloride.   The absorbance of the resulting colored

TABLE 11.2 Thin-Layer Chromatographic R_f Values and Visual Fluorescence Detection Limits of N-Nitrosamino Acids

	R_f			Detection limit[a]	
Compound	A	B	C	ng	pmoles
N-Nitrososarcosine	0.64, 0.70	0.71	0.75	20	180
Sarcosine	0.23	0.10[b]	0.02		
N-Nitroso-L-proline	0.63, 0.69	0.72	0.21	30	200
L-Proline	0.33	0.12[b]	0.03		
N-Nitroso-4-hydroxy-L-proline	0.60, 0.66	0.67	0.13	33	210
4-Hydroxy-L-proline	0.28	0.09[b]	0.01		

N-Nitrosamino acids are determined on activated silica gel by spotting, developing, irradiating with UV light, spraying with fluorescamine reagent, and viewing under longwave UV light. Amino acids are detected by spraying with ninhydrin reagent and heating. Solvent systems: A, 95% ethanol/benzene/water (4:1:1); B, methanol/chloroform (4:1); C, acetonitrile/chloroform/95% ethanol/acetic acid (100:100: 97:3).

[a] Determined in solvent system A.

[b] Streak.

Reprinted with permission from Ref. 13.

solution was measured at 550 nm. In the in situ fluorometric method, 7-chloro-4-nitrobenzo-2-oxa-1,3-diazole (NBD-Cl) was reacted with the amines, then silica gel chromatoplates were used to separate the fluorescent derivations. Immediately after development, the chromatoplates were air dried and then scanned with a Vitatron TLD flying spot densitometer in the fluorescence mode. The authors used successfully the colorimetric method in early work for the determination of N-nitrosamines in fried bacon and its cooked fat. However, they discovered that the color development reaction with N-1-naphthylethylenediamine dihydrochloride was very sensitive to the amount and nature of impurities in bacon. They thus examined the preparation of fluorescent amine derivatives with NBD-Cl, TLC

separation, and in situ fluorometric determinations. In a collaborative study, a common gas chromatography/mass spectroscopy method
for determination of N-nitrosamines was compared with both the colorimetric and in situ fluorometric methods. Comparable results were
obtained by all three approaches showing that less expensive alter-
nate methods were available for N-nitrosamines.

MYCOTOXINS

Stoloff (15) reviewed the analysis of mycotoxins by fluorodensitometry.
Fluorodensitometry has been employed extensively in the analysis
of mycotoxins. Mycotoxins are toxic metabolites of molds and may
be present as residues in food or feeds from uncontrolled growth
of certain undesirable mold species. In situ fluorometric methods
have been developed for such mycotoxins as aflatoxins, ochratoxins,
sterigmatocystin, zearalenone, and patulin. Several specific in
situ fluorometric methods for mycotoxins in food samples have been
published by the Association of Official Analytical Chemists (16).

Stack and Pohland (17) reported the results of a collaborative
study of a method for chemical confirmation of the identity of aflatoxins. The method was based on work by Przybylski (18). Aflatoxin
B_1 was converted to aflatoxin B_{2a} directly on a silica gel TLC chromatoplate. The test involved the catalytic action of trifluoroacetic
acid on the addition of water across the double bond of the terminal
furan ring of aflatoxin B_1 (17). Similarly, aflatoxin G_1 was converted to aflatoxin G_{2a}. Aflatoxin B_2 and aflatoxin G_2 were not
affected. In the procedure, 1 μl of trifluoroacetic acid was applied
directly on top of spots of an aflatoxin-containing extract and on
top of an aflatoxin standard on TLC chromatoplates. After a period
of time, the trifluoroacetic acid catalyst was evaporated, the chromatoplate developed, and then viewed under longwave ultraviolet radiation. Aflatoxin B_{2a} and aflatoxin G_{2a} appeared as blue fluorescent
derivatives. The chromatoplate was also sprayed with H_2SO_4 (1 + 3)
solution to change the aflatoxins' fluorescence from blue to yellow,

as additional confirmation for aflatoxins. In a collaborative study,
a total of 45 extracts, each derived from 9 g aflatoxin-free peanut
butter were prepared. Either 45 or 135 ng aflatoxin B_1 was added
to 36 of the extracts. This gave aflatoxin B_1 levels of 5 and 15
µg/kg peanut butter with the other 9 extracts designated as blanks.
The samples were randomly numbered and five samples each were distributed to nine collaborators along with 20 µg aflatoxin B_1 standards and instructions for carrying out the study. The collaborators
reported positive results using the trifluoroacetic acid test for
13 of the 16 samples to which aflatoxin had been added at the 5
µg/kg level and for all of the 17 samples to which 15 µg/kg had
been added. Identical results were obtained with the sulfuric acid
spray test. The results indicated that the test had good probability
of confirming as little as 5 µg aflatoxin B_1/kg sample. No positive
results were reported for any of the blank samples. The collaborators
commented that in most cases the methods were performed easily and
produced good results.

A collaborative study for the determination of zearalenone
in corn was reported by Shotwell, Goulden, and Bennett (19). The
study was conducted to decide whether a slightly modified method
developed by Eppley (20) for the screening of agricultural commodi-
ties for mycotoxins could be used to determine zearalenone in white
and yellow corn. Twenty-two laboratories participated in the study.
The samples were submitted to a series of separation steps, and
the final step was TLC. The zearalenone spots gave a greenish-blue
fluorescence under shortwave ultraviolet radiation. The chromatoplates were further sprayed with $AlCl_3$ solution and heated for 5
min at 130°C. Zearalenone gave a blue fluorescence. The amount
of zearalenone was determined by visual comparison with standards
or fluorodensitometrically. Statistical analysis of the data obtained
by the collaborators was given for the visual comparison method
and the densitometric method. The authors concluded that the slightly
modified Eppley method was applicable to the determination of zearalenone in corn and that it be considered as a possible official
method.

Tuinstra and Bronsgeest (21) developed an in situ fluorometric method for the screening and determination of aflatoxin M_1 in milk at the part per trillion level. The aflatoxin M_1 was isolated by extraction, column chromatography, and TLC. In the TLC step two-dimensional development was carried out. The aflatoxin M_1 was esti-mated visually or determined densitometrically. The limit of detec-tion for aflatoxin M_1 was 4 ng/kg; however, it was not stated whether the detection limit given was for the visual method or the densito-metric method.

Beljaars, Schumans, and Koken (22) discussed the quantitative fluorodensitometric determination of aflatoxins in nutmeg. They employed a modified analytical procedure consisting of an extrac-tion and column cleanup procedure of Scott and Kennedy (23). Also, a two-dimensional TLC procedure with silica gel chromatoplates de-scribed by Beljaars et al. (24) was used in the final separation step. A Vitatron TLC 100 was used to measure directly the reflected fluorescence of the aflatoxins on the chromatoplates. The authors also used a published test for the conformation of aflatoxins that involved hemiacetal formation. The TLC chromatographic and densito-metric procedure gave a coefficient of variation of 5.22 ± 1.24% with aflatoxin B_1. An average recovery of 92.6 ± 4.9% was obtained for aflatoxin B_1 spiked nutmeg samples. The coefficient of variation of the complete analytical procedure for ground nutmeg was 8.80%.

Salem and Swanson (25) described a fluorodensitometric method for patulin in apple products. Extraction, column chromatography, preparative TLC, and finally analytical TLC were used to isolate patulin from commercial samples such as frozen-concentrate apple juice and dehydrated apple pieces. Salem and Swanson enhanced the fluorescence of patulin adsorbed on silica gel chromatoplates by exposing the chromatoplates to concentrated ammonia fumes. The limit of detection for patulin was 0.1 μg on a chromatoplate and 1.0 mg of patulin per liter in apple juice. Salem and Swanson con-cluded their fluorodensitometric method required inexpensive instru-mentation, less laboratory time than other analytical methods, and could be adapted to routine analysis of patulin.

Athnasios and Kuhn (26) discussed an improved method for the isolation and estimation of sterigmatocystin in grains. Liquid-liquid extraction and TLC were used as separation steps. There was no need to use column chromatography, and that shortened the analysis time. Silica gel chromatoplates were employed in the final separation step, and the developed chromatoplates were sprayed with $AlCl_3$ reagent. Quantitation was accomplished by visually comparing the fluorescent intensity of the unknown spots with standard spots. Concentrations of sterigmatocystin as low as 50 µg/kg could be estimated. Their method was applied to barley, white and yellow corn, oats, and wheat. Percent recoveries of added toxin in the range of 50-400 µg/kg ranged from 83.6% for yellow corn and oats to >96% for barley, wheat, and white corn. Background interference was minimal for the method, and analysis time was reduced substantially by eliminating the need for a column cleanup step.

Buckley, Ikawa, and Sasner (27) reported the isolation of *Gonyaulax tamarensis* toxins from soft shell clams and developed a TLC fluorometric method for their detection and quantitation. The *Gonyaulax tamarensis* toxins are not mycotoxins but are mentioned here because they are very important in paralytic shellfish poisoning. Column chromatographic fractions from toxic clams were chromatographed on silica gel chromatoplates. The chromatoplates were sprayed with 1% hydrogen peroxide, heated and scanned with a Turner Model III fluorometer. With their method, 40-400 ng of the various poisons could be determined.

Recently studies have appeared with aflatoxins in which TLC methods with fluorodensitometric detection have been compared with high-performance liquid chromatography (HPLC) with ultraviolet and fluorescence detection (28,29). Pons (28) concluded that HPLC resolution of aflatoxins coupled with ultraviolet absorbance measurement is potentially a more accurate and precise analytical technique for the determination of aflatoxins B_1, B_2, and G_1 with sensitivity approaching present TLC methods. However, additional cleanup steps would be needed prior to HPLC to remove nonaflatoxin artifacts.

Seitz (29) investigated several HPLC columns and ultraviolet and
fluorescence detection for aflatoxins. He also used yellow corn
samples in his work. Seitz concluded that extensive studies will
be required to determine if HPLC provides better quantitation than
TLC methods. Also, these authors did not consider high-performance
thin-layer chromatography (HPTLC), which should improve the accuracy
and precision of in situ fluorometric methods. As with many analytical
approaches the choice of the approach depends on several things such
as time, cost, sensitivity and selectivity. In the future, both
HPLC methods and TLC methods should find wide use by analysts.

OXYTETRACYCLINE AND TETRACYCLINES

Willekens (30) developed a TLC method for the separation of the
impurities anhydrotetracycline, epianhydrotetracycline, epitetra-
cycline, and chlortetracycline from the antibiotic tetracycline.
Direct TLC fluorometry was used to determine the impurities.
Kieselguhr disodium ethylenediaminetetraacetate-impregnated chromatoplates were employed in the separation step, and a Zeiss TLC spectrophotometer was used in the fluorescence quantitation step. Willekens
presented an extensive table which listed accuracy and precision
data for the four impurities. The accuracy and precision were very
good, and Willekens concluded that direct TLC fluorometry is a very
sensitive way of evaluating the impurities in tetracycline. The
limit of detection was 50 to 100 times better than when spectrophotometry was used.

Later Willekens (31) reported a rapid and sensitive direct
TLC fluorometric method for evaluation of impurities in the antibiotic oxytetracycline. Impurities of interest were anhydrooxytetra-
cycline, epioxytetracycline, α-apooxytetracycline, and β-apooxytetra-
cycline. Again, Kieselguhr disodium ethylenediaminetetraacetate-
impregnated TLC chromatoplates were used in the separation step
and a Zeiss TLC spectrophotometer in the quantitation step. Accuracy
and precision were good even though curved calibration curves were

obtained. The method was sensitive to 10^{-2} µg of impurities in a
1% (w/v) solution of oxytetracycline.

A simple method for the quantitative analysis of tetracyclines
by direct fluorometry after TLC on cellulose chromatoplates was
developed by Ragazzi and Veronese (32). They observed that development of cellulose TLC chromatoplates with aqueous solutions of certain salts formed fluorescent tetracycline complexes. The salt solutions used were magnesium, calcium, barium, aluminium, and zinc
chloride at various concentrations in water. The samples investigated were tetracycline, 4-epitetracycline, anhydrotetracycline,
4-epianhydrotetracycline, oxytetracycline, chlorotetracycline, de-
methylchlorotetracycline, methacycline, doxycycline, and minocycline.
A Vitatron TLC flying spot densitometer was employed in the quantitative analysis experiments. The authors found good differentiation
among most of the tetracyclines when the various salt solutions
were used as mobile phases in the separation step. After development of the chromatoplates with the salt solutions, the chromatoplates
were viewed under ultraviolet radiation and the fluorescence was
noted. Ragazzi and Veronese studied the changes of fluorescence
with time and decided to measure the fluorescence of the tetracyclines
between 20 and 30 hr after the development of the chromatoplates.
They also investigated the influence of salt concentration of fluorescence, and the accuracy and precision of the method. The fluorescence calibration curves showed good linearity and the relative
standard deviation was good. For example, tetracycline gave a relative standard deviation of ±2.54% with calcium chloride solution as a
mobile phase. Magnesium chloride was suitable for the determination of
all the tetracyclines examined, and the other three salts produced
fluorescence that was useful for only some of the tetracyclines.

Szabo, Nagy, and Tomorkeny (33) described an in situ fluorometric
method for tetracyclines separated by TLC. A combination of metal
salts and organic solvents gave fluorogens with relatively high
fluorescence intensities. Cellulose was coated onto glass plates,
and prior to use the chromatoplates were submerged in a methanolic

ethylene glycol solution and then superfluous solvent was removed.
Samples were applied to the chromatoplate in three bands, and then
the chromatoplates were sprayed with magnesium chloride solution,
the second set with methanolic triethanolamine, and the third set
with magnesium chloride and then methanolic triethanolamine. The
chromatoplates were evaluated with a Camag Z scanner. The antibiotics
investigated were chlorotetracycline, α-doxycycline, oxytetracycline,
tetracycline, methacycline, β-doxycycline, and apoterramycin. The
linear range for the tetracyclines was from 0.05 to 1.00 µg.

Martin, Duncombe, and Shaw (34) described a thin-layer chromatographic test for tetracyclines and other compounds. The test was
based on the fluorescent or colored patterns developed by the thermal
degradation products of tetracyclines. For example, a sample was
spotted on a silica gel chromatoplate, the chromatoplate was heated
for 1 hr at 100°C, developed in an appropriate mobile phase, and
then viewed under longwave ultraviolet radiation. The authors proposed the approach for the identification of tetracyclines.

STEROIDS, PENICILLINS, CEPHALOSPORINS, SIOMYCINS, AND GENTAMICIN

Martin, Duncombe, and Shaw (34) applied the thin-layer chromatographic
test described in the previous section for the identification of
steroids, penicillins, and cephalosporins by degradation patterns
on chromatoplates. Steroids were spotted on fluorescent silica
gel chromatoplates, heated for 18 hr at 110°C, developed with a mobile
phase, then sprayed with an anhydrous zinc chloride solution, heated
for 20 min at 110°C, and finally the pattern of separated degradation
components was viewed under ultraviolet radiation. The test was
applied to 40 steroids and unequivocal identification was possible
in all cases. The test was also successfully applied to some for-
mulated preparations containing a steroid as the active agent.
The penicillins investigated were ampicillin, amoxycillin, benzathine
penicillin, benzylpenicillin, carbenicillin, cloxacillin, methicillin,
phenethicillin, phenoxymethylphenicillin, procaine penicillin and

propacillin. Ammonia solutions of the penicillins were spotted
on silica gel chromatoplates, heated at 150°C for 1.5 hr, developed
in a mobile phase, and examined under ultraviolet radiation. For
the cephalosporin antibiotics, cephradine, cephaloglycine, cephalexin,
cephazolin, cephalothin, and cephaloridine were separated by electrophoresis rather than by TLC. The Whatman paper was sprayed with
iodoplatinate reagent and then viewed under ultraviolet radiation.
The tests developed appeared to be suitable for general application
for the identification of complex organic materials and showed good
discrimination within the main classes of compounds studied.

Hirauchi and Masuda (35) developed a simple and rapid fluorodensitometric method for simultaneous determinations of siomycin
A, siomycin B, and siomycin D_1, in siomycins using silica gel-sintered
chromatoplates. After the chromatoplate was developed, it was sprayed
with 15 N sulfuric acid, then heated for 20 min at 105 ± 2°C. Within
80 min, the fluorescent spots of siomycin A, B, and D_1 were detected
with ultraviolet radiation and the fluorescence intensity measured
with a Hitachi MPF-2A spectrofluorometer equipped with a Hitachi
TLC scanning attachment. The linear range of a typical calibration
curve was 5-50 ng per spot. The method developed was 6 to 40 times
more sensitive than a previous method and also simpler.

Kabasakalian, Kalliney, and Magatti (36) described a method
for the determination of gentamicin complex components in fermentation broth by in situ fluorometric measurements of 4-chloro-7-nitro-
benzo-2-oxa-1,3-diazole (NBD chloride) derivatives. NBD chloride
reacts with primary and secondary amines. Gentamicin is a broad-
spectrum aminoglycoside antibiotic complex and is composed of three
components, C_1, C_2, and C_{1a}. They differ from each other in the
degree of methylation. The C_1, C_2, and C_{1a} components in broth
samples were separated with silica gel chromatoplates by a method
described by Wilson, Richard, and Hughes (37). The developed chromatoplates were dipped in methanolic NBD chloride for 2 sec, heated
at 120°C for 10 min, cooled, and rechromatographed in methanol in
the same direction as the first development. A Schoeffel SD 3000

spectrodensitometer was used to measure the fluorescence of the derivatives. Several experimental conditions were tested and the outline of the conditions described above was found to be the best. The main advantage the fluorometric method developed was the improved sensitivity compared to colorimetric methods.

A fluorescence densitometric method for the determination of gluconic and lactobionic acids ("sugar acids") in pharmaceutical preparations was described by Gübitz, Frei, and Bethke (38). Many methods are available for the "sugar acids," but most of these methods are nonspecific and lack sensitivity and precision for quantitation. The authors developed a method that did not have these disadvantages and permitted the determination of relatively low concentrations of "sugar acids" or their salts in pharmaceutical preparations. Silica gel chromatoplates were employed in the separation of the "sugar acids" from interfering sugars, citric acid and vitamin C. After drying the developed chromatoplate, a fluorescence reaction was carried out by dipping the chromatoplate quickly into a dipping solution. The dipping solution consisted of lead tetraacetate and 2,7-dichlorofluorescein in benzene. The reaction between the "sugar acid" and lead tetraacetate is probably based on a glycol cleavage of the "sugar acid" by lead tetraacetate. The dichlorofluorescein is converted irreversibly into a nonfluorescent oxidation product, which causes complete quenching of the background. However, on the spots containing the "sugar acids," part of the dichlorofluorescein remains undisturbed because the lead tetraacetate had reacted with the sugar acid. Thus some fluorescence remains from dichlorofluorescein and its intensity is proportional to the amount of "sugar acid." In the analysis of drug substances the limit of detection was 0.2 µg per spot, and using the data pair technique (39) the relative standard deviation was between 0.7 and 2.2%. For the analysis of tablets, the reproducibility was only slightly inferior to that of drug substances.

De Silva and Strojny (40), and de Silva, Strojny, and Stika (41) described methods in which the luminescence of fluorescamine derivatives of pharmaceuticals or the native luminescence of pharmaceuticals were observed on ethanol-saturated chromatoplates under a variety of conditions. For example, samples on chromatoplates were viewed at room temperature under shortwave and longwave ultraviolet radiation. In addition, the chromatoplates were submerged in liquid nitrogen and immediately after shutting off the ultraviolet radiation decaying phosphorescence signals were observed. Also, estimates were made on phosphorescent decay times. The main advantage of the above approach was the amount of luminescence information obtained for a given compound. The type of information obtained was very useful in characterizing pharmaceutical samples and could be used as a general screening procedure for pharmaceuticals.

Wisneski (42) developed a method for the determination of bergapten (5-methoxy-psoralen), a photoxin, in colognes, perfumes, and toilet waters. The bergapten was first separated from the sample by a series of extractions. The diluted extract was spotted on a silica gel chromatoplate and then two-dimensional TLC was carried out. The chromatoplate was dried and the emitted fluorescence measured with a Perkin-Elmer MPF-3 fluorescence spectrophotometer with a TLC accessory. The average percent recovery was 88% for levels of 0.001, 0.005, and 0.01% bergapten. The linear range for bergapten was from 0.2 to 1.0 μg.

Kouimtzis and Papadoyannis (43) reported a method for the determination of vitamins A, B_1, B_2, D_3, and C by in situ fluorometry. Silica gel chromatoplates were used in the separation step, and a Vitatron TLD 100 densitometer was employed to measure the fluorescence of the vitamins. The range of linearity varied for the different vitamins, but a typical range for vitamin D_3 was 0-10 μg. The detection limits were 30-70 ng for vitamins A, B_2, and D_3, and about 400 ng for B_1 and C. Vitamin A was scanned immediately because it was destroyed rapidly by aerial oxygen. The method was applied to polyvitamin syrup preparations and good accuracy was obtained, about 6%.

APPLICATIONS OF QUANTITATIVE HIGH-PERFORMANCE THIN-LAYER CHROMATOGRAPHY IN THE ANTIBIOTIC INDUSTRY

Kreuzig (44) discussed the application of high-performance thin-layer chromatography (HPTLC) in the antibiotic industry. Although he did not discuss solid surface luminescence analysis, his comments are pertinent because HPTLC can be used in conjunction with fluorescence and phosphorescence detection. Kreuzig stated that for inprocess control of antibiotic fermentation and for routine assays of samples, quantitative HPTLC can be used with advantage and is relatively inexpensive. In contrast to TLC, HPTLC has several advantages: (1) three times more samples can be applied on one chromatoplate; (2) separation time is 4 to 5 times faster; (3) very symmetrical spots are obtained. Kreuzig compared the analysis times for 12 samples of gramicidin using TLC and HPTLC and absorption measurements directly from the chromatoplates. The total time per sample with TLC was 22 min, and with HPTLC it was 8.5 min. A similar comparison was made for the analysis of ergosterol. The total time per sample with TLC was 10 min, and with HPTLC it was 2.5 min. The coefficient of variation for analysis on 10 different chromatoplates was 1.7-3.4% for gramicidin and 1.2-1.8% for ergosterol. There are obvious advantages to using HPTLC compared to TLC, and Kreuzig's comments and results can be extended to a variety of other samples. Also, in the future many applications should appear for HPTLC that use either fluorescence or phosphorescence in the identification step and in the quantitation step. Very few applications of this type have appeared in the literature.

REFERENCES

1. J. A. Yeransian, K. G. Sloman, and A. K. Foltz, *Anal. Chem.*, *51*, 105R (1979).

2. R. K. Gilpin, *Anal. Chem.*, *51*, 257R (1979).

3. B. C. Madsen and H. W. Latz, *J. Chromatogr.*, *50*, 288 (1970).

4. H. W. Latz and B. C. Madsen, *Anal. Chem.*, *41*, 1180 (1969).

5. J. H. Tatum and R. E. Berry, *J. Food Sci.*, *38*, 1244 (1973).

6. J. van Buren, L. de Vos, and W. Pilnik, *J. Food Sci.*, *38*, 656 (1973).

7. R. K. Ibrahim, *J. Chromatogr.*, *42*, 544 (1969).

8. E. R. Lieber and S. L. Taylor, *J. Chromatogr.*, *160*, 227 (1978).

9. E. R. Lieber and S. L. Taylor, *J. Chromatogr.*, *153*, 143 (1978).

10. E. J. Brach and B. Baum, *Appl. Spectrosc.*, *29*, 326 (1975).

11. V. W. Winkler, *Analytical Methods for Pesticides and Plant Growth Regulators*, Vol. X, G. Zweig and J. Sherma, eds., Academic Press, New York, 1978, pp. 545-559.

12. J. C. Young, *J. Chromatogr.*, *124*, 17 (1976).

13. J. C. Young, *J. Chromatogr.*, *151*, 215 (1978).

14. C. K. Cross, K. B. Bharucha, and G. M. Telling, *J. Agric. Food Chem.*, *26*, 657 (1978).

15. L. Stoloff, *Quantitative Thin Layer Chromatography*, J. C. Touchstone, ed., John Wiley & Sons, Inc., New York, 1973, Chap. 6.

16. *Official Methods of Analysis of the Association of Official Analytical Chemists*, 12th ed., Washington, D. C., 1975.

17. M. E. Stack and A. E. Pohland, *J. Assoc. Offic. Anal. Chem.*, *58*, 110 (1975).

18. W. Przybylski, *J. Assoc. Offic. Anal. Chem.*, *58*, 163 (1975).

19. O. L. Shotwell, M. L. Goulden, and G. A. Bennett, *J. Assoc. Offic. Anal. Chem.*, *59*, 666 (1976).

20. R. M. Eppley, *J. Assoc. Offic. Anal. Chem.*, *51*, 74 (1968).

21. L. G. M. Th. Tuinstra and J. M. Bronsgeest, *J. Chromatogr.*, *111*, 448 (1975).

22. P. R. Beljaars, J. C. H. M. Schumans, and P. J. Koken, *J. Assoc. Offic. Anal. Chem.*, *58*, 263 (1975).

23. P. M. Scott and B. P. C. Kennedy, *J. Assoc. Offic. Anal. Chem.*, *56*, 1452 (1973).

24. P. R. Beljaars, C. A. H. Verhülsdonk, W. E. Paulsch, and D. H. Liem, *J. Assoc. Offic. Anal. Chem.*, *56*, 1444 (1973).

25. T. F. Salem and B. G. Swanson, *J. Food Sci.*, *41*, 1237 (1976).

26. A. K. Athasios and G. O. Kuhn, *J. Assoc. Offic. Anal. Chem.*, *60*, 104 (1977).

27. L. J. Buckley, M. Ikawa, and J. J. Sasner, Jr., *J. Agric. Food Chem.*, *24*, 107 (1976).

28. W. A. Pons, *J. Assoc. Offic. Anal. Chem.*, *59*, 101 (1976).

29. L. M. Seitz, *J. Chromatogr.*, *104*, 81 (1975).

30. G. J. Willekens, *J. Pharm. Sci.*, *64*, 1681 (1975).

31. G. J. Willekens, *J. Pharm. Sci.*, *66*, 1419 (1977).

32. E. Ragazzi and G. Veronese, *J. Chromatogr.*, *132*, 105 (1977).

33. A. Szabo, M. K. Nagy, and E. Tomorkeny, *J. Chromatogr.*, *151*, 256 (1978).

34. J. L. Martin, R. E. Duncombe, and W. H. C. Shaw, *Analyst, 100*, 243 (1975).

35. K. Hirauchi and S. Masuda, *Chem. Pharm. Bull. (Tokyo)*, *25*, 1474 (1977).

36. P. Kabasakalian, S. Kalliney, and A. W. Magatti, *Anal. Chem.*, *49*, 953 (1977).

37. W. L. Wilson, G. Richard, and D. W. Highes, *J. Pharm. Sci.*, *62*, 282 (1973).

38. G. Gübitz, R. W. Frei, and H. Bethke, *J. Chromatogr.*, *117*, 337 (1976).

39. H. Bethke, W. Santi, and R. W. Frei, *J. Chromatogr. Sci.*, *12*, 392 (1974).

40. J. A. F. de Silva and N. Strojny, *Anal. Chem.*, *47*, 714 (1975).

41. J. A. F. de Silva, N. Strojny, and K. Stika, *Anal. Chem.*, *48*, 144 (1976).

42. H. H. Wisneski, *J. Assoc. Offic. Anal. Chem.*, *59*, 547 (1976).

43. Th. A. Kouimtzis and I. N. Papadoyannis, *Mikrochim. Acta, 145* (1979).

44. F. Kreuzig, *J. Chromatogr.*, *142*, 441 (1977).

12

Applications in Biochemistry, Medicine, and Clinical Chemistry

Solid surface luminescence analysis has been applied widely in biochemistry, medicine, and clinical chemistry, and several applications will be considered in this chapter. Solid surface fluorescence
analysis has found extensive use in the analysis of amino acids,
peptides, and proteins. In addition, a variety of components in
human body fluids such as drugs and pharmaceuticals have been analyzed
by solid surface fluorescence methods. Guilbault and co-workers
have developed several solid surface fluorescence methods for biolog-
ically important compounds. The application of room temperature
phosphorescence to compounds of clinical importance was already
considered in Chapter 7 and so will not be discussed in depth in
this chapter. The applications review in *Analytical Chemistry* is
an excellent source of information for the analysis of pharmaceuticals
and drugs by solution and solid surface luminescence analysis (1).
Also, Froehlich (2) published recently a review on the use of luminescence spectroscopy in clinical and biological analysis.

AMINO ACIDS, PEPTIDES, AND PROTEINS

Ragland, Pace, and Kemper (3) developed a very rapid and sensitive
fluorescence technique for detecting and quantitatively assaying
proteins in polyacrylamide gels. Proteins were labeled prior to

electrophoresis with fluorescamine and subjected to sodium dodecyl
sulfate (SDS) polyacrylamide gel electrophoresis. The three proteins
used in their study were sperm whale myoglobin, chymotrypsinogen
A from beef pancreas, and ovalbumin. After electrophoresis the gels
were placed in a quartz boat and scanned in a Gilford Model 2410-S
linear transport which was modified to measure fluorescence. Impurities were detected with fluorescamine in all three "purified" proteins but the labeled proteins separated into three distinct disks.
The fluorescence for a given amount of protein was very reproducible,
and the linear range was up to 7 µg of myoglobin, 9 µg of chymotrypsinogen A, and at least 12.5 µg of ovalbumin. The sensitivity was
about 0.5 µg for each protein and the precision was excellent. The
sensitivity was apparently dependent on the lysine content of the
protein. Also the sensitivity was as good as that observed with
other protein-staining methods for gels. The authors suggested
other potential applications for their system, namely, studies with
oligopeptides, and studies on protein structures.

The SDS electrophoresis on polyacrylamide gels of fluorescamine-
labeled proteins was also developed by Eng and Parkes (4). Protein
standards came from bovine serum albumin, ovalbumin, chymotrypsinogen,
lysozyme, myoglobin, and ribonuclease. Migration distances of the
protein-SDS complexes were measured under ultraviolet radiation
by observing the fluorescence of the fluorescamine-labeled proteins.
The authors showed there was a linear relationship between the logarithm of molecular weight of the proteins and migration distance.
Compared to conventional staining techniques with dyes such as Coomassie
brilliant blue, less time and smaller amounts of proteins were needed.
Also, smaller amounts of protein were needed compared to a dansyl-
labeled protein technique, and the fluorescamine-labeled products
were stable and could be stored frozen.

Eckhardt, Hayes, and Goldstein (5) developed a sensitive method
for the detection of glycoproteins after polyacrylamide gel electrophoresis. Dansyl hydrazine was condensed with aldehydic functions
generated by periodate oxidation of protein-bound carbohydrate.

The resulting hydrazones were reduced to stable hydrazine derivatives with sodium borohydride. After destaining, the fluorescent-labeled glycoproteins were visualized with ultraviolet radiation. The main advantage of the method was the increased sensitivity for detection of glycoproteins after polyacrylamide gel electrophoresis. Less than 40 ng of carbohydrate-bound *Bandeiraea simplicifolia* lectin was detected by the dansyl hydrazine technique.

A new fluorescent thiol reagent, N-(7-di-methylamino-4-methyl-coumarinyl) maleimide (DACM), that has a high quantum yield when reacted with protein SH groups, was described by Yamamoto, Sekine, and Kanaoka (6). The researchers used solution fluorescence in the determination of labeled proteins but detected labeled protein in polyacrylamide gel under ultraviolet radiation. The maximum emission wavelength of DACM was about 480 nm. Also, the quantum yield of DACM attached to denatured proteins was almost identical regardless of the protein.

Mendez and Lai (7) developed a general method for the detection of peptides and amino acids on paper with fluorescamine, after paper chromatography or electrophoresis. The application of fluorescamine to the visualization of peptides or amino acids on paper was not studied earlier systematically. The fluorescamine staining procedure involved heating paper at 55°C for 1 hr after electrophoresis or chromatography, washing with acetone, wetting with 1% triethylamine in acetone, drying at room temperature, washing with acetone, and drying. The fluorescent spots were detected by viewing under ultraviolet radiation. Secondary amines were converted to primary amines by treatment of the paper with acetic acid in acetone and N-chloro-succinimide in acetone. After washing and drying, the paper was treated as described above except a longer drying time in the final step was needed to enhance the fluorescence. The developed paper sheets or strips were photographed with a Polaroid M-P-3 Multipurpose Land Camera. The procedure was simple, rapid, and sensitive for the detection of amino acids and peptides. The limit of detection was approximately 50 pmoles of amino acid or peptide. This

was 5 to 20 times below that obtained by the conventional ninhydrin
procedure.

Fluorescence of tryptophan-containing peptides on paper or silica
gel after treatment with formaldehyde, formaldehyde-ozone or formaldehyde-hydrochloric acid was considered by Larsson, Sundler, and
Hakanson (8). Formaldehyde gas induced strong and characteristic
fluorescence from tryptophan and peptides with NH_2-terminal tryptophan residues on silica gel. The detection of small amounts of these
compounds on filter paper required the additional use of an oxidant
such as ozone. Treatment with formaldehyde-hydrochloric acid was
used for inducing fluorescence from tryptophan-containing peptides,
regardless of the position of the tryptophan residue in the peptide.
The spectral properties of the fluorophors of tryptophan-containing
peptides were distinctive and allowed distinction from all other
known biogenic compounds that were capable of giving fluorescence
with formaldehyde. Table 12.1 gives the minimum detectable amounts
of tryptophan, indoleacetic acid, and tryptophan-containing peptides
on filter paper and silica gel. Larsson, Sundler, and Hakanson
also studied the effect of hydrochloric acid, reaction time, and the
spectral characteristics of the fluorophors.

Mendez and Gavilanes (9) compared the fluorescence obtained
with o-phthalaldehyde (OPA) and fluorescamine in aqueous solutions
and on paper chromatograms. The detection of amino acids and peptides on paper chromatograms with fluorescamine was reported by
Mendez and Lai (7) and was discussed earlier in this section. The
procedure for detection with OPA on paper after electrophoresis or
chromatography was to dry the paper at 50°C for 1 hr, wash the paper
with acetone, dip the paper in a tray containing a solution of triethylamine and 2-mercaptoethanol in acetone, dry the paper, dip the
paper in a tray containing an OPA solution, wash the paper with
acetone, and finally, dry the paper at room temperature. The fluorescent spots were detected with ultraviolet radiation and photographed.
Mendez and Gavilanes found that amino acids were more readily visualized at the 500 pmole level with OPA than with fluorescamine,

TABLE 12.1 Minimum Detectable Amounts (µg) of Tryptophan, Indoleacetic Acid and Tryptophan-containing Peptides

Compound	Filter-paper			Silica gel		
	formaldehyde	formaldehyde-ozone	formaldehyde-HCl	formaldehyde	formaldehyde-ozone	formaldehyde-HCl
L-tryptophan	0.3-1.0	0.03	0.03	0.03	0.1	0.03
Indoleacetic acid	-	-	0.03	-	-	0.03
L-tryptophyl-L-alanine	0.3-1.0	0.03	0.03	0.03	0.1	0.03
L-tryptophyl-L-glutamic acid	0.3-1.0	0.03	0.03	0.1	0.1	0.1
L-tryptophyl-L-glycine	0.3-1.0	0.03	0.03	0.1	0.1	0.1
L-tryptophyl-L-phenylalanine	0.3-1.0	0.03	0.03	0.1	0.1	0.1
L-tryptophyl-L-tyrosine	0.3-1.0	0.03	0.03	0.1	0.1	0.1
L-glycyl-L-tryptophan	-	-	0.03	-	-	0.1
L-phenylalanyl-L-tryptophan	-	-	0.03	-	-	0.1
L-prolyl-L-tryptophan	-	-	0.03	-	-	0.1
L-argenyl-L-tryptophyl-L-glycine	-	-	0.1	-	-	0.3
Tetragastrin	-	0.1	0.3	0.1-0.3	0.3	0.3
Peptavlon	-	-	0.1-0.3	-	-	0.3
Experimental allergic encephalitogenic peptide	-	-	0.3-1.0	-	-	0.3
Glucagon	-	-	a	-	-	-
Thyreoglobulin	-	-	-	-	-	-

TABLE 12.1 (Continued)

Volumes of 1 µl were applied. Results indicated by dashes (-) represent amounts greater than 1 µg.

[a] With 10 µg of glucagon, weak fluorescence was observed.

Reprinted with permission from Ref. 8.

but peptides were more easily detected with fluorescamine. The fluorescent spots produced by fluorescamine remained visible for several hours. They also found a greater sensitivity for detection of peptides with fluorescamine on thin-layer chromatograms.

Reisfeld and Levi (10) developed ultramicrodetermination methods for amino acids and histidyl-peptides by fluorescence measurements of spots on paper. Whatman 120 filter paper was found to give the most homogeneous spots. Fluorophors were obtained by the reaction of OPA with histidine and histidyl-peptides, and other fluorophors were obtained by the reaction of amino acids with ninhydrin and phenyl-acetaldehyde. The linear fluorescence range was 2-400 ng for histidine and histidyl-peptides with OPA reagent, and linear fluorescence range was 7-250 ng for glycine, phenylalanine, histidine, and methionine with ninhydrin-phenylacetaldehyde reagent. The precision of the method for the histidine-OPA fluorophor was less than 2%. The limit of detection was 2 ng for the OPA reaction and 4 ng for the ninhydrin-phenylacetaldehyde reaction.

Sherma and Touchstone (11) presented calibration curves for several amino acids to illustrate the quantitation of amino acids by in situ fluorescence densitometry on chromatoplates. Alanine, glycine, aspartic acid, phenylalanine, and tyrosine were studied. Silica gel and cellulose chromatoplates were tested as solid surfaces. After development of the chromatoplates, they were placed in a hood until the mobile phase was evaporated. The chromatoplate was then sprayed with a 10% solution of triethylamine in methylene chloride, fluorescamine reagent, and again with the triethylamine. The purpose of the second triethylamine spray was to stabilize the fluorescence (12). A Schoeffel SD 3000 spectrodensitometer was used in

the reflectance mode to measure the fluorescence. All calibration
curves were linear to at least 0.5 µg. The detection limits varied
from 10 ng for aspartic acid on silica gel to 150 ng for aspartic
acid on cellulose. Generally the detection limits were about 20-30
ng. The authors proposed their approach be used for determining
amino acids in urine, blood plasma, or serum. With their methods,
direct application of small samples seemed quite possible.

Felix and Jimenez (12) first reported the method of spraying
a chromatoplate with a solution of triethylamine before and after
spraying the chromatoplate with a fluorescamine solution to stabilize
and enhance the fluorescence. This approach was used by Sherma and
Touchstone (11) as discussed previously. Felix and Jimenez tested
the stability of the fluorescent spots of 5.0 nmoles each of L-alanine,
L-glutamic acid, L-lysine, and L-alanyl-L-alanine on silica gel chromatoplates. The fluroescence intensities of the model compounds were
faded only slightly after 24 hr. The authors stated that the presence
of triethylamine neutralized any residual acetic acid from the developing system and stabilized the fluorophor in the triethylammonium
salt form, thus preventing conversion to the nonfluorescent γ-lactone.
Picomole amounts of the model amino acid and peptide were detectable
visually.

Detection of fluorescamine-labeled amino acids, peptides, and
other primary amines on silica gel chromatoplates was discussed
by Imai et al. (13). Instead of spraying chromatoplates with fluorescamine after separation of amino acids and peptides, the authors
formed fluorescamine derivatives in solution first, then carried
out TLC separation. The approach improved the sensitivity compared
to previous methods. The fluorescamine derivatives of primary amines
were detectable at levels below 100 pmoles on chromatoplates when
viewed under ultraviolet radiation. With many compounds as little
as 10 pmoles could be detected. Some of the compounds investigated
were aspartic acid, leucine, arginine, L-alanyl-L-alanine, and gly-
cylproline. The authors emphasized how the use of fluorescamine
for prelabeling of primary amines offered distinct advantages over

the commonly used dansyl procedure even though the fluorescence properties and intensities of the fluorescamine and the dansyl fluorophors
were comparable. One problem with dansylation arises from the presence
of the fluorescent reaction byproducts, dansyl-acid and dansyl-amide.
Ordinarily they must be removed by tedious extraction procedures.
In addition to amino groups, dansyl chloride reacts at other sites
such as the hydroxyl group of tyrosine or the imidazole nitrogen
of histidine. Because of the side reactions, more than one fluorophor is produced from an amine. With fluorescamine, each peptide,
catecholamine, and amino acid gave one product.

Vladovska-Yukhnovska, Ivanov, and Malgrand (14) developed a
fluorescence method for the detection of hexosamine (amino sugars)
derivatives on silica gel chromatoplates. They prepared Dis-chloride
(diphenylindenone sulfonyl chloride) derivatives of glucosamine,
galactosamine, and several amino acids, and then separated the
derivatives by TLC. After the separation step, the chromatoplate
was observed under ultraviolet radiation and the Dis derivatives
gave a yellow-green fluorescence. The fluorescence method allowed
the detection of amino sugars in amounts as low as 2×10^{-11} mole.

Segura and Gotto (15) developed a procedure for the detection
and quantitation of a variety of organic compounds on several different
ferent chromatoplates. Fluorescent derivatives were formed by thermal
treatment of the chromatoplates in the presence of ammonium hydrogen
carbonate. They listed the excitation and emission maxima for glycine, serine, phenylalanine, tyrosine, tryptophan, and methionine
adsorbed on silica gel chromatoplates. Generally microgram amounts
of material could be detected.

Nakamura and Pisano (16) developed a method for derivatization
of compounds at the origin of chromatoplates with fluorescamine.
The compounds were dissolved in buffer and were applied to TLC plates
and then derivatized at the origin by developing with or dipping
into an acetone-hexane solution of fluorescamine. Practically all
fluorescamine derivatives stayed at the origin, and they were separated
subsequently using appropriate mobile phases. The method was applied

to numerous peptides, amino acids, and amines. The detection limit
was near 10 pmoles. Nakamura and Pisano commented that prelabeling
methods in which the sample is derivatized before spotting on a chromatoplate have two shortcomings: (1) dilution of the samples with
reagents, and (2) the need to react each sample individually. Nakamura
and Pisano's method was more sensitive than the method developed
by Imai et al. (13) discussed earlier. Nakamura and Pisano (17)
expanded on their work and emphasized that acidic mobile phases could
not be used to develop fluorescamine derivatives because the fluorescamine derivatives were not stable in acidic media. Nakamura and
Pisano did not comment on the types of acids nor the concentration
of the acids. They did employ acetic acid in the mobile phases.
Compounds first separated by TLC with acidic mobile phases have been
detected with fluorescamine sprays, but the sensitivity of the spraying method amounts to only a fraction of the theoretical sensitivity,
because of incomplete reaction and high background. Nakamura and
Pisano (17) detected compounds with primary amine groups by a dipping method in which the developed chromatoplate was dipped into
a fluorescamine solution and then observed under ultraviolet radiation. Their approach gave high reactivity, and peptides, proteins,
and most amines of biological interest were detected at the 10 pmole
level.

Touchstone et al. (18) considered the optimal use of fluorescamine for in situ thin-layer chromatographic quantitation of amino
acids. Alanine, glycine, aspartic acid, phenylalanine, and tyrosine
were used as model compounds. Silica gel chromatoplates were employed
in the separation step and a Schoeffel SD 3000 spectrodensitometer
was used in the measurement step. Comparisons were made with different drying techniques and spraying techniques. Drying at 110°C
for 10 min resulted in poor results for some amino acids. Larger
fluorescence intensities were obtained when the chromatoplates were
dried at room temperature for 30 min. Spraying of the fluorescamine
in dimethylsulfoxide or dimethylformamide proved to be suitable
because no triethylamine stabilization was required. They found

the reaction time was dependent, regardless of the spraying method used. The emission intensity appeared to develop a high intensity within 30 min, followed by a rapid decline. The fluorescence intensity increased to a maximum at 3 hr, which was followed by a second decline in intensity. The authors did not discuss detection limits.

Lindeberg (19) reported on the use of OPA for the detection of amino acids and peptides on thin-layer chromatograms. Lindeberg pointed out that the popular reagent, fluorescamine, is expensive and precludes its everyday use in many laboratories. OPA in the presence of a strong reducing agent such as 2-mercaptoethanol produces highly fluorescent compounds with most amino acids (20). After development of silica gel or cellulose chromatoplates, the chromatoplates were sprayed with a solution of OPA and 2-mercaptoethanol and 5 min later with a triethylamine solution. After 10 min the chromatoplates were viewed under ultraviolet radiation. Table 12.2 gives the minimum quantities of some amino acids and peptides detected. The difference in stability on the two adsorbents was important. On silica gel the spots decayed within a few minutes at room temperature, while on cellulose the spots were very stable even when heated. Lindeberg concluded OPA reagent was comparable to fluorescamine in sensitivity and convenience and had the additional advantage of being considerably less expensive.

Nakamura (21) described the TLC detection of histidine, histamine, and histidyl peptides at picomole level using a fluorogenic reaction with fluroescamine. The approach was based on a reaction in which the compounds were derivatized with fluorescamine, converted into different fluorescent products by heating in strong acid medium, and then separated on silica gel chromatoplates with an appropriate mobile phase. As little as 4-60 picomoles of the histidine derivative could be detected under ultraviolet radiation, and other classes of compounds gave no fluorescent spots. OPA has gained wide acceptance as a sensitive chromatographic reagent for histidine, histamine, and histidyl peptides. The OPA methods are the most sensitive for

TABLE 12.2 Minimum Quantities of Some Amino Acids and Peptides Detected

	Detection limit (pmoles per spot)								After TLC
	cellulose				silica gel				
Substance	10 min	1 hr	2 hr	1 hr (100°)	10 min	1 hr	2 hr	1 hr (100°)	10 min
Alanine	250	250	250	250	50	250	500	1000	100
Arginine	10	10	25	25	10	50	100	500	25
Asparagine	250	250	250	250	25	100	250	500	50
Aspartic acid	100	100	100	100	25	100	100	1000	100
Cystine	10	10	10	10	100	250	250	100	250
Glutamine	100	100	100	100	25	100	250	1000	50
Glutamic acid	100	100	100	100	25	100	250	1000	100
Glycine	100	100	100	100	25	250	1000	---	100
Histidine	25	50	50	50	25	100	100	100	25
Isoleucine	100	100	100	100	50	500	500	---	100
Leucine	100	100	100	100	50	500	500	---	100
Lysine	25	25	25	25	25	100	250	1000	50
Methionine	100	100	100	100	50	100	250	1000	100
Phenylalanine	50	50	50	50	50	250	500	1000	100
Proline	---	---	---	250	---	---	---	250	250[a]
Serine	100	100	100	100	10	250	250	1000	25

Threonine	100	100	100	100	25	250	250	1000	50
Tryptophan	25	25	25	50	50	100	100	100	250
Tyrosine	25	25	25	50	50	500	500	500	100
Valine	25	25	25	50	50	500	1000	---	100
Glycyl-glycine	250	250	250	250	100	500	1000	---	250
Bradykinin	50	50	50	50	50	500	1000	250	100
Arginine-vasopressin	100	100	100	500	250	1000	1000	1000	500

1-μl aliquots containing 5-1000 pmoles were spotted on thin-layer plates, dried at 100°C for 30 min, treated with the spray reagents and viewed under UV light (350 nm) after the time indicated. Thin-layer chromatography (TLC) was carried out on silica gel with n-butanol-acetic acid-water (4:1:1); front migration, ca. 10 cm; dried and visualized as above.

[a] After heating at 100° for 1 hr.

Reprinted with permission from Ref. 19.

chromatographic detection of imidazole compounds; however, they
lack specificity because OPA reacts with compounds having a primary
amino group and compounds other than imidazole compounds. The method
developed by Nakamura is both sensitive and highly specific for
histidine, histamine, and histidyl peptides. A positive response
was restricted to 2-(4-imidazolyl)-ethylamines having the following
structures:

ANALYSIS OF DRUGS AND PHARMACEUTICALS IN HUMAN BODY FLUIDS AND ANIMAL TISSUE

A special issue of the *Journal of Chromatographic Science* (May 1974)
covered the analysis of drugs of abuse. An extensive bibliography
of references on the analysis of drugs of abuse from 1972 to 1974
appeared at the end of the volume. Many references were given for
the in situ fluorometric detection and quantitation of drugs and
pharmaceuticals. Several articles were given that included discus-
sions on fluorescence detection and measurement of drugs and pharmaceuticals from solid surfaces. Jain and Cravey (22) considered
the identification of barbiturates from biological specimens. Paper
chromatography and TLC methods were reviewed along with colorimetric
and fluorometric detection methods. The identification of nonbar-
biturate hypnotics from biological specimens was presented by Cravey
and Jain (23). TLC, colorimetric, and fluorometric visualization
methods were employed. Mule (24) discussed methods for the analysis
of morphine and related surrogates. Some of the methods involved
TLC and fluorescence detection. Bastos and Hoffman (25) compared
methods for the detection of amphetamines, cocaine, and metabolites.
The compositions of mobile phases for the silica gel TLC of amphetamines
and cocaine were given along with fluorescence detection methods
and other detection methods. Speaker (26) emphasized the importance

of TLC and visualization methods in analytical toxicology laboratories. Many of the visualization reagents produce fluorescent derivatives on chromatoplates.

A spectrofluorodensitometric method for flurazepam and its
major metabolites in blood was developed by de Silva, Berkersky,
and Publisi (27). Flurazepam has been used in the treatment of
insomnia. The method involved the selective extraction of flurazepam
and its major metabolites into ether and hydrolysis in acid to their
respective benzophenones. The benzophenones were cyclized in di-
methylformamide-potassium carbonate to their respective 9-acridanone
derivatives and then separated by TLC with silica gel chromatoplates.
The fluorescence of the 9-acridanones was measured in the reflection
mode with a spectrodensitometer. The limits of detection were 0.5-1.0
ng. Blood levels of flurazepam following 30 ng therapeutic doses
are barely detectable by solution fluorescence after dilution; however, they can be quantitated with much greater precision by direct
scanning fluorometry. The authors presented data obtained from
blood samples of two human subjects taken over a 24 hr period showing
that the method was readily adaptable to clinical situations.

Only a few methods have been developed for the simultaneous
determination of the antidepressant, amitriptyline, and its major
metabolite, nortriptyline. Farber, Mulder, and Man In 'T Veld (28)
developed an in situ fluorometric method for the simultaneous determination of these two compounds. Both drugs were isolated from
plasma by a single extraction step. Part of the organic phase was
evaporated to dryness and the residue was dissolved in ethanol containing hydrochloric acid. TLC was then used to separate amitriptyline and nortriptyline and their metabolites. Amitriptyline and
its metabolites were rendered fluorescent by immersing the chromatoplate in dilute perchloric acid and subsequently heating the chromatoplate in a drying oven. The fluorescence was measured with a
Vitatron TLD-100 densitometer. The concentration of the unknown
was interpolated between two reference standards. The recovery
of amitriptyline from plasma was 104%, and 81% for nortriptyline.

The standard deviation for each compound was 8%, and the sensitivity of the method was about 10 ng/ml. The method was both sensitive and specific for therapy control purposes and was also inexpensive. About 3 hr were needed for the simultaneous analysis. The approach was very specific for the drugs, and only a few hypnotics and tranquilizers interfered. However, this problem could be minimized by using a different mobile phase in the TLC step.

Farber and Man In 'T Veld (29) used essentially the same procedure described in the preceding paragraph to determine the anticonvulsant, carbamazepine, in blood. The average recovery was just over 100% and the limit of detection of the method was about 0.5 µg/ml in plasma. The metabolites of carbamazepine, 10,11-dihydroxy-carbamazepine and carbamazepine-10,11-epoxide, were separated from carbamazepine in the TLC step. The R_f values of a number of other medicines were investigated and only three had almost the same R_f value as carbamazepine, but their fluorescence was very low and did not interfere in the determination.

Simmons and DeAngelis (30) developed in situ methods for pyrimethamine in plasma and urine. Pyrimethamine, a 2,4-diaminopyrimidine, has been used for its antimalarial activity. 2,4-Diaminopyrimidines have been used in the treatment of meningeal leukemia. Simmons and DeAngelis found that NH_4HSO_4 enhanced the fluorescence of pyrimethamine and other 2,4-diaminopyrimidines on silica gel chromatoplates. Once generated, the fluorescence product was stable for several weeks. Extraction and TLC steps were used to isolate pyrimethamine from plasma and urine. After the chromatoplate was developed, it was sprayed evenly with NH_4HSO_4 solution, dried, and then the fluorescence was measured with a Schoeffel SD 3000 spectrodensitometer. As little as 5 ng could be measured on a chromatoplate. A recovery of 99.4 ± 7.5% was obtained for pyrimethamine in urine for the drug concentration of 0.3-5.0 µg/ml and 98.1 ± 4.8% in plasma for the drug concentration of 0.2-1.25 µg/ml.

Van Hoff and Heyndrickx (31) described the analysis of amphetamine and amphetamine analogs after reaction with 4-chloro-7-nitrobenzo-2,1,3-oxadiazole (NBD-Cl). Extraction and TLC procedures were used to isolate the compounds from urine and blood samples. Silica gel chromatoplates were employed in the TLC procedures and an Aminco-Bowman spectrophotofluorometer equipped with a thin-film scanner was used in the fluorescence measurement step. The fundamental reaction underlying the determination was the coupling of primary and secondary alkylamine functions with NBD-Cl. All the fluorescent derivatives had a maximum fluorescence between 510 and 545 nm (Table 12.3). The method allowed for rapid identification of the most important amphetamine analogs and no interferences from nicotine or other coextractives were encountered. The method was more sensitive than existing TLC detection methods and allowed determination of therapeutic amphetamine levels in 1-ml blood samples.

TABLE 12.3 Maxima in Fluorescence Spectra of NBD-Derivatives

Compound	Maximum in fluorescence spectrum (nm)
Amphetamine	523
Pervitine	537
Methoxyphenamine	535
Ethylamphetamine	530
Ritaline	535
Lidepran	535
Preludine	535
Sympatol	525
Effortil	545
Chlorphentermine	535
Vasculat	525
β-Phenylethylamine	540
Ephedrine	525
Heptaminol	512

Reproduced from Ref. 32 with the permission of the copyright owner.

A specific TLC tissue residue determination of sulfadiazine
following fluorescamine derivatization was reported by Sigel, Woolley,
and Nichol (32). Sulfadiazine has been used as an antimicrobial
agent in humans and for treatment of infections in food-producing
animals. The authors actually gave procedures for the determination
of sulfadiazine in milk, muscle, eggs, liver, fat, skin, kidneys,
and plasma. The final separation step involved TLC with a silica
gel chromatoplate. The developed chromatoplate was dipped once
quickly in a fluorescamine solution to derivatize the sulfadiazine.
A Schoeffel SD 3000 was used to measure the fluorescence from the
chromatoplate. Their method for sulfadiazine had the advantages
that the sample preparation was simpler and the time for analysis
was shorter compared to colorimetric and GLC methods. About 8 to
10 tissue samples could be assayed by one person in 1 day. The
TLC method could be adapted for the determination of sulfonamide
drug mixtures.

There has been considerable interest in the role of polyamines
in the control of the growth process in animals. Abe and Samejima
(33) developed a fluorometric method for the determination of spermidine and spermine in tissues by TLC. Many present methods are inadequate in that they lack simplicity, sensitivity, reproducibility,
and accuracy. Abe and Samejima used N-3-aminopropylheptane-1,7-
diamine as an internal standard and added the internal standard to
homogenized rat tissue samples. After separation of the internal
standard, spermidine, and the spermine from a tissue sample, a concentrate was spotted on a silica gel sintered-glass plate and developed
with an appropriate mobile phase. After development of the chromatoplate, it was developed again with a fluorescamine solution, dried,
and the fluorescence intensity measured with a fluorometer equipped
with a TLC scanner. No diffusion or tailing of spots was observed
with the fluorescamine treatment step. The fluorescence of the
spots changed with time; however, peak area ratios of spermidine
and spermine, with respect to the internal standard, were almost
constant during the time period examined. The relative standard

deviations on the same chromatoplate were in the range of 1.7-3.8%
for spermidine and 1.1-4.1% for spermine. The use of an internal
standard increased the reliability of their method. The internal
standard had three important properties: (1) its basicity is similar
to spermidine and spermine, (2) it was located between spermidine
and putrescine on a chromatoplate so that putrescine did not interfere with its detection, (3) it had almost the same pattern as spermidine and spermine in developing and fading properties of fluorescence
on chromatoplates. Also, the authors' procedure for development
with fluorescamine solution allowed for uniform covering of the
chromatoplates with fluorescamine, with minimum consumption of the
reagent. The limit of detection of the method was about 100 pmoles.

    Fleisher and Russell (34) prepared 5-dimethylaminoaphthalene-1-
sulfonyl (Dns) derivatives of ammonia, 1,3-diaminopropane, 1,4-di-
aminobutane, 1,5-diaminopentane, spermidine, histamine, and spermine,
and separated the derivatives by TLC with silica gel chromatoplates.
They were interested in diamines and polyamines because elevated
levels of these compounds occur in the urine of patients with metas-
tatic cancer. The urine samples analyzed by Fleisher and Russell
were always accompanied by a sample containing 5 nmoles each of putrescine, cadaverine, spermidine, and spermine. This sample was
subjected to the complete analytical procedure, and the amine concentration of the experimental samples was estimated by the height
of the corresponding peaks in the standard sample. The accuracy
and precision of the method were good. The sensitivity of the dansylation method was better than that of an amino acid analyzer method
and could be used for very small sample volumes.

    Heby and Andersson (35) reported a simplified micromethod for
the quantitative analysis of putrescine, spermidine, and spermine
in urine. The method had a minimal number of preparative steps and
allowed for the analysis of a large number of clinical samples.
After acid hydrolysis of urine, the polyamines were converted to
fluorescent 1-dimethylamino-naphthalene-5-sulfonyl (Dns) derivatives
and separated on silica gel chromatoplates with an appropriate mobile

phase. The developed chromatoplate was sprayed with a triethanol-
amine-propan-2-ol solution and then dried for 16 hr. An Aminco-
Bowman spectrophotofluorometer equipped with a TLC scanner was used
to measure fluorescence. Reduced sample preparation was accomplished
by omitting several concentration steps and the extraction of polyamines into 1-butanol or isoamyl alcohol. Recoveries of polyamines
added to urine samples were 98% for putrescine, 95% for spermidine,
and 89% for spermine. The Dns derivatives of putrescine, spermidine,
and spermine appeared as well-defined and well-separated spots, and
linear calibration curves were obtained between 25 and 500 pmole
per spot. The coefficient of variation for the analysis of mixtures
of the polyamines was about 10%. The polyamine concentrations were
determined successfully in urine samples from normal patients and
pateints with a variety of disease entities.

Farber, De Kok, and Brinkman (36) described a method for the
determination of digitoxin in human serum. Digitoxin is prescribed
in cases of cardiac failure. Older procedures for digitoxin are
very time-consuming or not specific enough for digitoxin. The researchers isolated digitoxin from serum by a single extraction with
chloroform and a concentrated residue was spotted on a Keiselgel
60 chromatoplate and developed with chloroform/methanol/acetone/
water (64:6:28:2). The TLC system separated digitoxin from its
metabolites and digoxin. The fluorescence of the spots was generated
by treatment with hydrogen chloride vapor under the influence of
a quartz-halogen lamp, and a Vitatron TLC 100 densitometer was used
to measure fluorescence. The concentration of the sample was obtained
by interpolation between two reference standards. Extensive recovery
studies in the therapeutic range of 10-50 ng/ml showed an overall
recovery of 99.1% with a standard deviation of 11.2%. The limit
of detection was about 1 ng on a chromatoplate and about 24 determinations could be made in 1 day. The method was not too expensive,
had good accuracy and precision, and was specific and sensitive
for the determination of digitoxin in the serum of patients in the
therapeutic and toxic range of 10-60 ng/ml.

A thin-layer fluorescence procedure with a sensitivity limit
of 0.8 ng/ml for triprolidine, an antihistamine agent, was developed
by DeAngelis, Kearney, and Welch (37). The method was used for quan-
titating triprolidine in plasma following oral administration of a
therapeutic dose to human subjects. Plasma samples were extracted
and a concentrated extract was spotted on a silica gel chromatoplate
and developed. The developed chromatoplate was sprayed with a solution of ammonium bisulfate and dried for about 1 hr, and the fluorescence was measured with a Schoeffel SD 3000 spectrodensitometer.
The exact nature of the fluorophor formed on the chromatoplate after
spraying triprolidine with ammonium bisulfate is unknown, but it
is relatively stable to ultraviolet radiation. It was possible
to quantitate as little as 0.4 ng of triprolidine on a chromatoplate.
Plasma triprolidine levels were determined successfully in the plasma
samples of several male subjects.

Straughan et al. (38) quantitated the tranquilizer, chlordiazepoxide, and its four major metabolites in biological fluids by measurement of fluorescence from chromatoplates. Spectrofluorometric and
gas-chromatographic methods for the determination of chlordiazepoxide
and its metabolites are often time-consuming, do not determine all
metabolites, are often affected by interfering substances, and may
have inadequate sensitivity. The authors have reported an accurate,
precise, and simple method for the simultaneous assay of chlordiazepoxide and its four major metabolites in biological fluids using
quantitative TLC. Following extraction from serum and urine samples,
chlordiazepoxide and its four major metabolites were separated by
TLC. Prazepam was added as an internal standard to the serum and
urine samples. The developed chromatoplate was placed for 45-50
min in a glass tank containing a small beaker of red, fuming nitric
acid. The chromatoplate was removed from the tank, left under a
fumehood for 30 min, and then placed in an oven at 100°C for 30
min. This procedure enhanced the fluorescence of the spots considerably. The fluorescence of the spots was measured with a Schoeffel
SD 3000 spectrodensitometer, and a linear-linear plot of the benzo-
diazepine-prazepam peak-height ratio versus concentration was made.

The peak-height ratio of each benzodiazepine in the unknown was compared to its respective standard curve for quantitation. The recovery of each of the five compounds added to serum and urine samples was 80% or greater. The coefficient of variance was between 1.4 and 6.4%, and the standard curves for all five compounds were linear from at least 0.1 to 2.0 μg. The limit of detection for all five compounds was between 0.05 and 0.1 μg. A wide spectrum of potentially interfering drugs were tested and found not to interfere with the determinations.

A quantitative TLC method was developed for procainamide and its major metabolite, N-acetylprocainamide, in plasma. Procainamide is a cardiac depressant. Gupta, Eng, and Lewis (39) developed the method which involved extraction of plasma with dichloromethane to isolate the drug and its metabolite. p-Amino-N-(2-dipropylamino-ethyl)-benzamide and its N-acetyl derivative were added to plasma prior to extraction as internal standards. A TLC solvent system was used which gave excellent separation of the four compounds from each other, from plasma constituents, and from other drugs prescribed commonly with procainamide. A Zeiss TLC scanning spectrophotometer was used to measure the fluorescence of the components separated on silica gel chromatoplates. The percent recoveries varied from 88 to 98%, and the relative standard deviations for procainamide and N-acetylprocainamide were ±2.7% and ±1.2%, respectively. The method was very sensitive and there was no problem distinguishing plasma containing 0.5 mg/liter of procainamide or N-acetylprocain-amide from a blank plasma. Also, as little as 50 μl of plasma could be assayed. The authors commented on several existing methods for the drug and its metabolite. Colorimetric methods lack specificity, gas chromatographic methods are more specific but are slow, and high performance liquid chromatographic methods are also relatively slow. Gupta, Eng, and Lewis further commented that with the increasing demands for drug analysis, they opted to use TLC techniques wherever possible.

Terbutaline is a stimulant used widely in the treatment of bronchial asthma and reversible bronchospasm. Tripp et al. (40)

developed a method for the determination of the intact drug and
total terbutaline in human serum. Metaporterenol was added as an
internal standard to serum samples. Terbutaline and metaproterenol
were isolated from serum by ion-exchange fractionation then extracted
into tert-butyl alcohol. The compounds were further separated by
TLC, and fluorescence was induced by heating with glyoxylic acid.
The fluorescence was measured in the reflectance mode with a Schoeffel
SD 3000 spectrodensitometer. Peak-height ratio, terbutaline to
metaproterenol, was plotted versus nanogram per milliliter terbutaline
for the calibration curves. For the determination of total terbutaline, samples were hydrolyzed overnight at 37°C with arylsul-
fataseglucuronidase. For the determination of intact terbutaline,
the hydrolysis step was omitted. Twelve assay samples could be ac-
commodated on a single chromatoplate. Concentrations as low as
1 ng/ml could be measured, and the overall recovery percentage of
added terbutaline was 102 ± 12%. The method was adopted readily
in the analysis of blood samples of nine male volunteers. The approach offered an alternative method for a reported gas chromatography/
mass spectrometric method and allowed for the determination of total
terbutaline.

Kupke and Zeugner (41) described a quantitative high-performance
thin-layer chromatographic (HPTLC) method for lipids in plasma and
liver homogenates after direct application of 0.5-μl samples to
silica gel chromatoplates. Lipid extraction prior to TLC is time-
consuming, thus Kupke and Zeugner were interested in a direct application procedure. The HPTLC procedure separated neutral and polar
lipids in capillary blood plasma and liver homogenates. A total
of 14 samples could be applied to one HPTLC plate. Fluorescence
was induced by a procedure developed by Segura and Gotto (15) which
was described earlier in the chapter, and a Zeiss TLC scanning spectrophotometer was used to measure the fluorescence. Table 12.4
shows the responses of reference lipids and the precision of fluorescence measurement. The recovery of known amounts of lipid was
96-100%, and a linear relationship between peak area and amount
of lipid was found in the nmole range. The plasma lipids of patients

TABLE 12.4 Relative Responses of Lipids and Precision of Fluorescence Measurement

Lipid	Amount (nmole per spot)	Peak area (cm^2)	cm^2/nmole	C.V. (%)
Cholesterol	1.04	6.26 ± 0.13	6.02	2.1
Cholesterol esters	2.71	15.74 ± 0.23	5.82	1.5
Triacylglycerols	1.07	7.84 ± 0.22	7.33	2.8
Phosphatidylcholine	1.6	8.7 ± 0.2	5.4	2.3

The lipids of Lipid-Trol (a standard reference sample) were separated by HPTLC and the peak areas were taken as the product of peak height and half-width of the peak. The coefficients of variation (C.V., %) were calculated from 14 samples per HPTLC plate.

Reprinted with permission from Ref. 41.

suffering from various illnesses and of healthy patients were analyzed using reference methods and HPTLC. Cholesterol esters, triacylglycerols, and phosphatidylcholine gave identical values. Slightly lower (7%) values were obtained with the HPTLC method for cholesterol.

Van der Merwe and Steyn (42) reported a TLC method for the determination of diazepam and its major metabolite, N-desmethyldiazepam, in human serum. The method was developed to replace a time-consuming hydrolysis-GC method in a bioavailability study. Samples were extracted, and the concentrated extract was spotted on silica gel chromatoplates and developed. The chromatoplate was dried, exposed to HCl gas, and then irradiated with 254 nm radiation. The procedure converted diazepam, the metabolite, and an internal standard, carbamazepine, to fluorescent compounds. The fluorescence was enhanced by dipping the chromatoplates into 10% paraffin wax in petroleum ether. Percent recoveries from spiked serum samples ranged from 102 to 105% for diazepam and from 100 to 102% for the metabolite.

The fluorescence photometric determination of disopyramide and its metabolite, mono-N-dealkylated disopyramide, in plasma after separation by TLC was reported by Gupta et al. (43). The method involved extracting plasma samples, spotting the concentrated extracts

on silica gel chromatoplates, and developing the chromatoplates
in an appropriate mobile phase. The developed chromatoplates were
dipped in 20% sulfuric acid in methanol to induce the fluorescence
of the spots. A Zeiss TLC scanning spectrophotometer was used to
measure fluorescence. The authors found better overall precision
without the use of an internal standard. The accuracy and precision
of the method were good, and the compounds could be quantitated
down to a level of 0.5 mg/liter of plasma. The procedure was more
specific than a solution fluorometric method and allowed for the
simultaneous determination of the parent drug and its metabolite.
Also, present gas-chromatographic procedures require complex and
tedious sample preseparation steps and are not suitable for routine
laboratory analysis.

The fluorometric determination of atenolol in plasma and urine
by direct evaluation of thin-layer chromatograms was reported by
Schafer and Mutschler (44). Atenolol is a cardioselective beta-
blocking drug. The authors commented there are one solution fluorometric method and two gas chromatographic methods for atenolol in
plasma and urine samples. The detection limit for the solution
fluorometric method is 50 ng/ml and for the two gas chromatographic
methods about 10-20 ng/ml. With the gas chromatographic methods,
atenolol is converted to heptafluorobutyric anhydride. Schafer
and Mutschler were able to obtain a detection limit of about 5 ng/ml
with a relatively simple method. For plasma samples, extraction and
TLC separation were employed, and for urine samples just TLC was
used. The developed chromatoplates were sprayed with citric acid
solution. The fluorescence was determined on the moist chromatoplates because the fluorescence decreased rapidly as the chromatoplate dried. A Zeiss TLC scanning spectrophotometer was employed
to measure the fluorescence. Table 12.5 gives the mean recoveries
and relative standard deviations for atenolol determinations in plasma
and urine. No explanation was given for the relatively low-percentage
recovery for atenolol in plasma. The authors commented that 10%
atenolol is excreted in the urine in the form of hydroxyatenolol,

TABLE 12.5   Means of Recoveries and Relative Standard Deviations
for Atenolol Determinations in Plasma and Urine[a]

Sample	Concentration of atenolol	Recovery (%)	Relative standard deviation (%)
Plasma	500 ng/ml	65.5	3.78
	100 ng/ml	62.9	5.98
	20 ng/ml	66.0	12.06
Urine	100 µg/ml	102.4	4.77
	20 µg/ml	96.7	1.56
	2 µg/ml	85.4	5.43

[a] Eight samples of each of three different concentrations were investigated.

Reprinted with permission from Ref. 44.

and their experiments with TLC indicated that interference by hydroxyatenolol in the determination of atenolol was improbable.

LOW-TEMPERATURE PHOSPHORESCENCE FROM SOLID SURFACES

Szent-Gyorgyi (45) suggested detecting substances on paper chromatograms at liquid nitrogen temperature by their fluorescence or phosphorescence in ultraviolet radiation. This technique apparently
has been found to be of little use in the past. Gordon and South
(46) published a short paper describing the detection of the fluorescence of compounds adsorbed on paper at liquid nitrogen temperature.
They pointed out that the technique was usually not applicable when
the paper had been exposed to ultraviolet absorbing solvents such
as phenol or pyridine. Compounds investigated were phenylpyruvic
acid, 2,8-dihydroxy-6-methylaminopurine, L-tyrosine, benzoic acid,
and L-phenylalanine.

Randerath (47) described the detection of purine derivatives
on poly(ethyleneimine) -- or PEI -- cellulose ion-exchange chromatoplates in the nanogram range by phosphorescence at 77°K. The chromatoplate was treated as follows. About 200 ml of liquid nitrogen

were allowed to evaporate in a metal tray, which was placed on a thick layer of insulating plastic material. After the tray was precooled, the chromatoplate was put immediately in the tray and its center was held down with a bent spatula. Metal weights were placed at the corners of the sheet. While the center of the sheet was held down, liquid nitrogen was poured into the tray beside the sheet so that it formed a layer approximately 1 cm deep on top of the chromatoplate. After a few seconds, the spatula was removed from the center of the chromatoplate. The chromatoplate was then illuminated briefly with ultraviolet radiation, and the phosphorescent afterglow was photographed. Randerath concluded in 1967 that PEI-cellulose TLC and low-temperature phosphorescence appeared to be the most sensitive methods for unlabeled adenine and quanine derivatives available. Mayer, Holman, and Bridges (48) commented that Randerath did not emphasize visual detection of low-temperature phosphorescence. They described the visual phosphorescence detection of purine-containing compounds on cellulose chromatoplates. The minimum detectable limit for adenosine was 0.75 nmole and adenine 1.00 nmole. The authors emphasized that the approach was an extremely sensitive, nondestructive, in situ method for purines and there was no interference from fluorescent contaminants because of the short-lived lifetimes of fluorescence.

Isenberg et al. (49) used the intrinsic phosphorescence of unstained proteins to locate their position in polyacrylamide gels. Impurities in the gel caused no serious problems. Upon removal of the exciting lamp, the background emission decayed rapidly and each protein spot or band continued to glow for many seconds thereafter. The intrinsic phosphorescence of proteins originates from the tyrosines and tryptophans. Photographs showing the phosphorescence were presented for gels with separated components from beef-heart lactate dehydrogenase, ribosomal proteins, and calf-thymus histones.

O'Donnell and Winefordner (50) reviewed advances in instrumentation and methodology in phosphorimetry for clinical analysis, including a brief discussion of room temperature phosphorescence.

(See Chapter 7 for a further discussion of room temperature phosphorescence.) O'Donnell and Winefordner mentioned a windowless
sample cell for low-temperature measurement for front-surface excitation measurement of emission. The system was useful for samples
such as solids, slurries, and pastes that cannot be handled by conventional commercial phosphorescence cells. Spectra of various
phosphors were similar to those obtained by conventional methods.
Detection limits in the nanogram range were given for atropine, caf-
feine, p-nitrophenol, phenylalanine, quinine, tryptophan, and yohim-
bine.

De Silva, Strojny, and Stika (51) characterized pharmaceuticals
adsorbed on thin-layer chromatoplates by low-temperature fluorescence
and phosphorescence. Their results were discussed in Chapter 11.

WORK OF GUILBAULT AND CO-WORKERS

Guilbault and co-workers have developed methods for the assay of
enzymes, substrates, activators, and inhibitors using solid surface
fluorescence analysis. The instrumentation and techniques for their
methods were described in Chapter 3. In this section several of
the methods they developed will be discussed. Lau and Guilbault
(52) and Guilbault (53-55) have reviewed the assay of clinically
important components by solid surface fluorescence analysis.

Guilbault and Zimmerman (56) reported the first successful
measurement of the rate of an enzyme reaction on a solid surface
using fluorescence detection. Cholinesterase (ChE) was determined
in horse serum. ChE is important because individuals with anemia,
malnutrition, and pesticide poisoning have low levels of this material. High levels of ChE indicate a nephrotic syndrome. For the
determination of ChE, pads were prepared by placing 10 µl of a 10^{-2} M
solution of N-methylindoxyl acetate on a silicone rubber pad and
evaporating to dryness (see Chapter 3). To start the determination,
20 µl of a sample enzyme solution was applied to the pad. When
the pad was placed in the light bean of a fluorometer, a recorder

was started immediately and the rate of change of fluorescence
measured. From a calibration curve of the change in fluorescence
units per minute versus enzyme concentration, the activity of ChE
in the sample solution was obtained.

$$\text{N-methylindoxyl acetate} \xrightarrow{\text{ChE}} \text{N-methyl indoxyl (fluorescent)}$$

The rate of production of N-methylindoxyl was proportional to the
ChE concentration. Calibration curves were obtained from 10^{-6} to
10^{-2} units/ml, and the precision and accuracy of the method were
2% and 2.2%, respectively. Studies indicated that pads were stable
for at least 30 days if kept in a cold, dark place.

The determination of human serum alkaline phosphatase (AP)
was reported by Guilbault and Vaughan (57). High levels of alkaline
phosphatase are obtained in rickets, Paget's disease, obstructive
jaundice, and metatastic carcinoma. Naphthol AS-BI phosphate was
deposited on the surface of a silicone rubber pad and aliquots of
buffer solution and serum were placed on the pad. The drops were
mixed and spread into a uniform shape. The pad was fastened onto
a blackened metal plate and was placed on a cell holder of an Aminco-
Bowman filter fluorometer. The rate of the reaction was recorded,
and the serum AP value was obtained from the initial rate of the
reaction.

$$\text{Naphthol AS-BI phasphate} \xrightarrow{\text{AP}} \text{naphthol AS-BI (fluorescent)}$$

The method could be applied to all serum samples except those which
were highly jaundiced. The parameters pH, incubation time, drop
volume, size and shape; and the shape and color of the pads were
discussed. Table 12.6 shows the results from a series of serum
samples. Over 80 serum samples were analyzed and a relative error
of ±5% was obtained.

Zimmerman and Guilbault (58) developed a fluorescence method
for the determination of lactate dehydrogenase (LDH). This enzyme
is elevated in individuals with acute and chronic leukemia in re-
lapse, myocardial infactions and carcinomatosis.

$$\text{Lithium lactate + NAD} \xrightarrow{\text{LDH}} \text{NADH + pyruvic acid (fluorescent)}$$

TABLE 12.6   Determination of Serum Alkaline Phosphatase

Units alkaline phosphatase taken	Units alkaline phosphatase found[a,b,c]
20	21.5
22	19
24	24.5
26	23
28	25
30	31
31	31
33	37
37	34
39	40
41	39
43	42
48	49
53	54[d]
61	60[d]
64	64[d]

[a] Units expressed as μmole phenolphthalein liberated $min^{-1} l^{-1}$ serum.

[b] Obtained by multiplying the rate by 31.5.

[c] Average from three or more different samples -- each analysis in duplicate.

[d] One sample -- average of two determinations.

Reprinted with permission from Ref. 57.

The determination of serum LDH was based on the above enzyme catalyzed reaction. Calibration curves were plotted in terms of the change in fluorescence units per minutes versus enzyme concentrations. In this way, the rate of production of NADH fluorescence was equated to LDH activity. In the procedure, a 50-μl aliquot of NAD solution was applied to a pad and evaporated over silica gel under reduced

pressure to produce a thin film of solid NAD on the surface of the
pad. Then a 20-µl aliquot of the lactate solution was added to
the pad and evaporated in a similar manner which produced a thin
film of solid lactate. The pad could be stored over silica gel at
atmospheric pressure and no noticeable decomposition of the reactant
film was noted over a period of a month. A 10-µl aliquot of glycine
buffer with semicarbazine hydrochloride was added to the pad and
spread over the pad so the entire substrate film was dissolved in
the drop of solution. Then the pad was placed into a fluorometer
and the fluorescence background rate was recorded. After recording
the background fluorescence, a 20-µl aliquot of the serum sample
was put on the pad. The pad was placed immediately into a fluorometer
and the fluorescence rate recorded. From 160 to 1000 units of LDH
in serum was determined with an average relative error of 2.3%.
For a given sample the complete analysis took only 3-5 min (52).

The measurement of creatine phosphokinase (CPK) in serum is
valuable in the diagnosis of mycordial and skeletal muscular dystrophy.
Lau and Guilbault (59) described a "reagentless" fluorometric method
for the analysis of serum CPK activity. The method was based on
the use of silicone rubber pads upon which all the reagents were
placed for CPK determination. The rate of formation of NADH fluorescence was measured and related to CPK activity. The following reactions indicate the chemistry involved:

$$\text{Creatin phosphate} + \text{ADP} \xrightarrow{\text{CPK}} \text{creatine} + \text{ATP}$$

$$\text{ATP} + \text{glucose} \xrightarrow{\text{hexokinase}} \text{ADP} + \text{glucose-6-phosphate}$$

$$\text{Glucose-6-phosphate} + \text{NAD} \xrightarrow{\text{G-6-PDH}} \text{6-phosphogluconate} + \text{NADH} + \text{H}^+$$

All the reagents and enzymes for an assay were immobilized in lyophilized form on the surface of a pad. The sample was added and
then the fluorescence was measured. Lau and Guilbault used the
Worthington Statzyme CPK kit which contained all the enzymes, cofactors,
and substrates for CPK assay. The advantages of their approach were
low cost, speed, no external reagents, stability of reagent pads,
selectivity, and small sample size (3-25 µl).

Kuan, Lau, and Guilbault (60) developed an enzymatic fluorometric method for serum urea based on the rate of disappearance of NADH fluorescence. They also reported a simplified pad-preparation method that provided uniformity, low background and little contamination.

$$\text{Urea} + H_2O \xrightarrow{\text{urease}} 2NH_4^+ + CO_2$$

$$NADH + NH_4^+ + \alpha\text{-ketoglutarate} \xrightarrow[\text{dehydrogenase}]{\text{glutamate}} NAD^+ + \text{glutamate} + H_2O$$

All the reagents for the determination of urea were lyophilized on the surface of a silicone rubber pad. The serum sample was simply placed on the pad and the fluorescence measured for an assay. Table 12.7 shows some typical percent recovery data for urea. They found a within-day precision of 2.7% (coefficient of variation), and the same serum assayed on each of 3 days gave a coefficient of variation of 3.8%. They compared their method with results obtained by the diacetyl monoxime method performed with a Hycel Mark X multi-channel analyzer and the coefficient of correlation was 0.998. The method was much faster than when following conventional procedures. Only 5 min per assay was needed once the prepared pad was brought to room temperature.

TABLE 12.7   Recovery Study

Urea added (mg/dl)	Total urea found	Recovery (%)[a]
0	11.3	–
5.0	16.2	98
7.5	19.1	104
10.0	21.3	100
0	10.6	–
5.0	15.6	100
7.5	18.3	102
10.0	20.2	96

[a] Average of three determinations.

Reprinted with permission from Ref. 60.

Rietz and Guilbault (61) discussed the fluorometric assay of serum
α-glutamyltransferase in solution and on silicon rubber pads. γ-Glutamyltransferase is normally present in serum and increased activity in
serum indicates heptobiliary or pancreatic disease. The following reaction formed the basis for the method.

N-γ-L-glutamyl-α-naphthylamide + glycylglycine
$$\xrightarrow[\text{pH 9.0}]{\text{γ-glytamyltransferase}}$$ α-naphthylamine (fluorescent)
+ L-γ-glutamylglycylglycine

Solution and solid surface measurements were performed at 37°C. Measurements on silicone rubber pads were made after lyophilizing a stable
substrate film on a pad. Only 10-50 µl of buffered substrate and serum
containing the enzyme to make a total volume of 60 µl were necessary
for each assay. The rate of appearance of α-naphthylamine fluorescence,
liberated from N-γ-L-glytamyl-α-naphthylamide by the enzymatic action
of γ-glutamyltransferase, was measured and equated to its activity in
serum. Pads were stored at -20°C in the dark, in sealed bottles containing desiccant. The pads were stable under the storage conditions
and could be kept for about a month with little deterioration. For an
assay one only needed to add serum and buffer to a pad that had been
brought to 37°C and then measure the fluorescence. Both the solution
and solid surface fluorescence methods had good precision and accuracy.

Rietz and Guilbault (62) reported a solution and solid surface fluorometric assay of serum acid or alkaline phosphatase. The procedures
were similar to those described above. In Paget's disease, patients
have very high serum-alkaline phosphatase activity, and a moderate to
great increase in activity indicates obstructive jaundice. Serum acid
phosphatase activity is increased in prostatic cancer. Rietz and
Guilbault's method was based on the formation of fluorescent 4-methyl-
umbelliferone. 4-Methylumbelliferone was liberated from 4-methylumbel-
liferone phosphate by enzymatic action.

4-methylumbelliferone phosphate $\xrightarrow{\text{acid or alkaline}\atop\text{phosphatase}}$

4-methylumbelliferone + PO_4^{3-}

Results indicated the fluorometric solution and solid surface methods
agreed well with those obtained by accepted spectrophotometric methods.
The coefficients of variation showed that the solid surface fluores-

cence methods were more precise than the solution fluorescence methods.
The percentage recovery was very good for both approaches.

The determination of glucose in blood samples by solid surface
fluorescence was developed By Kiang, Kuan, and Guilbault (63). The
method was based on the sequential coupling reactions with flucose-6-
phosphate dehydrogenase and $NADP^+$. The rate of appearance of NADPH
fluorescence was detected and related to the glucose concentration in
plasma or serum. The glucose concnetration in 50 fresh plasma samples
was determined by their method and by the o-toluidine spectrophotometric
method and a correlation coefficient of 0.986 was obtained. The accuracy and precision of the method were very good, and the main advantages were low cost and rapid response. Table 12.8 shows the
percentage recovery for glucose from plasma and serum samples.

Guilbault (54) has also discussed fluorometric methods for
creatine in urine, uric acid in serum, and cholesterol determination. There are several advantages to the silicone rubber pad methods
developed by Guilbault and co-workers. Since essentially all the
reagents are present on the pads, there is no need for further reagent
preparation. The pad method is a microprocedure and thus very small
amounts of reagent are needed. Because fluorescence procedures are
used, the approach is several degrees of magnitude more sensitive
than colorimetric methods. In conventional methods, temperature

TABLE 12.8 Recovery Study

Plasma or serum glucose concentration (g/liter)	Glucose (g/liter)		Recovery (%)
	added	found	
0.91	0.50	1.42	102
0.91	1.00	1.92	101
1.42	0.50	1.90	96
0.88	0.50	1.37	98
0.88	1.00	1.91	103
0.88	1.50	2.34	97

Reprinted with permission from Ref. 63.

must be controlled in all enzyme assays based on a kinetic approach.
The fluorescent pad method is temperature independent. Provided
the sample is at the same temperature as used to prepare a calibration curve, the temperature of the environment has little effect
on the results. Many substrates and enzyme solutions used for present
clinical procedures are unstable and new solutions must be prepared
fresh daily. When reagents are placed in solid form on the silicone
rubber matrix, they can be kept for weeks or months (52).

REMAINING APPLICATIONS

In this section examples are discussed that did not fit readily
into the previous sections. For example, work with biochemical
standards that were not identified or determined in body fluids will
be considered.

Sperling (64) used thin-layer chromatography and gas chromatography to identify D-lysergic acid diethylamide (LSD). Two mobile
phase systems were employed to separate LSD and 21 other compounds.
The separated alkaloids were viewed under ultraviolet radiation,
and they fluoresced blue. Gas-chromatographic conditions were also
discussed.

Ranieri and McLaughlin (65) described the use of fluorescamine
as a thin-layer chromatographic visualization reagent for alkaloids.
They were interested in distinguishing between primary and secondary
amines with no interfering reactions from other functional groups
such as phenols or imidazoles. The fluorescamine spray did not
interfere with the subsequent use of additional spray reagents.
For example, a developed chromatoplate containing alkaloid extracts
was first sprayed with fluorescamine to visualize primary amines.
Then spraying the chromatoplate with 5-dimethylaminoaphthalene-1-
sulfonyl chloride (Dns-Cl) showed fluorescent derivatives of phenols
and imidazoles and converted secondary amines from fluorescamine
conjugates (dark purple) to fluorescent Dns conjugates (yellow).
Finally spraying with iodoplatinate allowed visualization of ter-

tiary amines. The authors tested 34 cactus alkaloids and related compounds.

The quantitative fluorodensitometric determination of ergot alkaloids was discussed by Prosek et al. (66). The authors presented the conditions needed to obtain useful fluorescence signals in the reflectance mode. Cellulose chromatoplates were used to separate the alkaloids, and calibration curves and reproducibility data were given. In a later paper, the same authors discussed the possibility for direct total assay of ergot alkaloids after separation by TLC using silica gel chromatoplates (67). The chromatoplates were activated for 30 min at 105°C and impregnated in a 60:35:5 (v/v) mixture of absolute ethanol/formamide/5% ammonia. The chromatoplates were then dried and heated for 2.5 min at 105°C. These chromatoplates were used to separate the alkaloids. The alkaloids were separated efficiently when an impregnating solvent that contained 30-40% formamide was used. Also, very strong fluorescence was obtained from alkaloids with chromatoplates impregnated with a solvent that contained 30-40% formamide. Conditions for densitometry on high-performance TLC chromatoplates were given, and results obtained by conventional TLC, high-performance TLC, liquid chromatography, and Van Urk color reaction were compared.

Karlsson and Peter (68) developed a method for the determination of alkaloids from *Lupinus polyphyllus* that was simpler, more rapid, and more sensitive than procedures in which Dragendorff reagent is used. The alkaloids were extracted with methanol from previously ground leaves of *Lupinus polyphyllus* and separated on silica gel chromatoplates. After development, the chromatoplates were dried and heated for 17 hr at 130°C in a drying chamber. The heat treatment induced strong fluorescence from the alkaloids on the chromatoplates (15). The detection limit was about 10 ng and a standard deviation of ±4% was achieved. Quantitative results were given for alkaloids from a bitter type and a sweet type of *Lupinus polyphyllus*. Results were given for 13-hydroxylupanine, angustifoline, lupinine, lupanine, and sparteine.

Ichimasa, Ichimasa, and Uranaka (69) reported the fluorometric
determination of lipids on chromatoplates sprayed with 2',7'-dichloro-
fluorescein. Work was done with lipid standards and separations
were carried out on silica gel chromatoplates. The developed chromatoplates were dried and then uniformly sprayed with an 0.02% 2',7'-
dichlorofluorescein solution. The fluorescence was measured with
a Shimazu CS 900 dual-wavelength TLC scanner with an attachment
for fluorometry. Calibration curves were given for some neutral
lipids, phospholipids, and cholesterol.

Vinson and Hooyman (70) discussed a sensitive fluorogenic visualization reagent for the detection of lipids on silica gel chromatoplates. The ammonium salt of 8-anilino-1-naphthalenesulfonate (ANS)
was used as a spray reagent. The results from ANS were compared
with the results from other spray reagents, namely, phosphomolybdic
acid, 2',7'-dichlorofluorescein, and rhodamine B. After separation
of the lipids, the chromatoplates were sprayed with the appropriate
reagent. The neutral lipids, cholesterol, oleic acid, triolein,
and cholesteryl, gave lower detection limits with ANS reagent compared to the reagents dichlorofluorescein and rhodamine B. The
phospholipids, sphingomyelin and lecithin, yielded lower detection
limits with ANS than with phosphomolybdic acid. Detection limits
for both neutral lipids and phospholipids were in the subnanogram
range.

Sherma, Dobbins, and Touchstone (71) described the quantitation
of morphine by fluorescence, fluorescence quenching, and color development using in situ spectrodensitometry after separation with
silica gel chromatoplates. Also, fluorescence was used for in situ
quantitation of amphetamine separated by TLC after reaction with
fluorescamine. Calibration curves were linear in all four cases
up to 1 μg. Morphine was detectable at 25 ng with the fluorescence
method and at 100 ng with the quenching and visible color methods.
Amphetamine was detected at 100 ng after spraying with fluorescamine.
The fluorescence procedure was recommended because of its simplicity
and superior sensitivity. The authors concluded the methods would

probably be applicable to drugs isolated initially from urine by
any frequently used separation methods.

Thapliyal and Maddocks (72) reported the TLC separation of azathioprine and 6-mercaptopurine metabolites as phenyl mercury derivatives and detection of the derivatives by low-temperature (-196°C)
fluorescence. Azathioprine is used as an immunosuppressive drug particularly in patients with organ transplants. Phenyl mercury derivatives were prepared to protect the thiol group from oxidation.
Silica gel chromatoplates with a fluorescent indicator were used
in the experiment. At low temperature, under ultraviolet radiation
at 254 nm, the derivatives were mainly blue or dark spots against
the fluorescent background of the chromatoplate. At 366 nm the
spots fluoresced weakly, but the fluorescence became intense after
spraying with 2 N HCl. As little as 1.5 ng 6-mercaptopurine could
be detected visually.

Cashion et al. (73) briefly described the detection of low
levels of ultraviolet-absorbing fluorescence-quenching compounds
on paper chromatograms. They superimposed a fluorescent screen
on the paper chromatogram while examining the paper chromatogram
with an ultraviolet light from underneath. They used polyethylene-
backed thin-layer sheets containing silica gel impregnated with
a fluorescent indicator as the fluorescent screen. Nucleosides
and nucleotides were detected at the nanomole and subnanomole levels.

Nakai et al. (74) described the fluorescence detection of glycolaldehyde on silica gel chromatoplates after spraying the chromatoplates with a solution of o-aminodiphenyl. Glycolaldehyde exists
in the solid state as a symmetrical dimer and in solution as an
equilibrium mixture of the monomer, the symmetrical dimer, and an
unsymmetrical dimer. Apparently because of the monomer and dimers,
glycolaldehyde is usually detected as double or triple spots on
chromatoplates sprayed with color reagents. With o-aminodiphenyl
the monomer of glycolaldehyde showed the fluorescence reaction but
not the dimer. Because of this the authors were able to detect
the monomeric aldehyde. Several monosaccharides were chromatographed,

and most of these sugars remained at or near the origin of the chromatoplate, and none of the sugars overlapped glycolaldehyde on the chromatoplate. The limit of visual detection was found to be 0.05 µg.

Sabatka et al. (75) reported three chromatographic systems for
the separation of digoxin and dihydrodigoxin. TLC with silica gel,
TLC with cellulose, and paper chromatography were used. Also five
different spray reagents were employed. A reagent containing ascorbic
acid, methanol, hydrochloric acid, and hydrogen peroxide was the
most sensitive fluorometric reagent and gave a 1-ng limit of detection for digoxin and a 10 ng limit of detection for dihydrodigoxin.

Taylor, Blass, and Ho (76) described the thin-layer chromatographic
separation and spectrofluorometric determination of lithocholic acid.
Lithocholic acid, a bile acid, has been isolated from human bile,
feces, serum, and urine. In the authors' work, lithocholic acid
was separated in 5 sec from other bile acids, i.e., chenodeoxycholic,
deoxycholic, and cholic acid. A Seprachrom miniature chromatography
chamber was employed. The developed chromatoplate was dried, and
the chromatoplate was sprayed with 5% H_2SO_4 in methanol and then
placed in an oven at 95°C for 5 min. The fluorescence intensity
of lithocholic acid was measured with a Farrand Mark I spectrofluorometer equipped with a thin-layer scanning attachment. A linear
relationship was obtained for lithocholic acid within the range
25-175 ng, and the coefficient of variation of a 150-ng standard
was 4.2%.

REFERENCES

1. R. K. Gilpin, *Anal. Chem.*, *51*, 257R (1979).

2. P. Froehlich, *Appl. Spectrosc. Reviews,* *12*, 83 (1976).

3. W. L. Ragland, J. L. Pace, and D. L. Kemper, *Anal. Biochem.*,
    59, 24 (1974).

4. P. R. Eng and C. O. Parkes, *Anal. Biochem.*, *59*, 323 (1974).

5. A. E. Eckhardt, C. E. Hayes, and I. J. Goldstein, *Anal. Biochem.*,
    73, 192 (1976).

6. K. Yamamoto, T. Sekine, and Y. Kanaoka, *Anal. Biochem.*, *79*,
    83 (1977).

7. E. Mendez and C. Y. Lai, *Anal. Biochem.*, *65*, 281 (1975).

8. L. I. Larsson, F. Sundler, and R. Hakanson, *J. Chromatogr.*, *117*, 355 (1976).

9. E. Mendez and J. G. Gavilanes, *Anal. Biochem.*, *72*, 473 (1976).

10. R. Reisfeld and S. Levi, *Anal. Lett.*, *10*, 483 (1977).

11. J. Sherma and J. C. Touchstone, *Anal. Lett.*, *7*, 279 (1974).

12. A. M. Felix and M. H. Jimenez, *J. Chromatogr.*, *89*, 361 (1974).

13. K. Imai, P. Bohlen, S. Stein, and S. Udenfriend, *Arch. Biochem. Biophys.*, *161*, 161 (1974).

14. Y. Vladovska-Yukhnovska, Ch. P. Ivanov, and M. Malgrand, *J. Chromatogr.*, *90*, 181 (1974).

15. R. Segura and A. M. Gotto, Jr., *J. Chromatogr.*, *99*, 643 (1974).

16. H. Nakamura and J. J. Pisano, *J. Chromatogr.*, *121*, 33 (1976).

17. H. Nakamura and J. J. Pisano, *J. Chromatogr.*, *121*, 79 (1976).

18. J. C. Touchstone, J. Sherma, M. F. Dobbins, and G. R. Hansen, *J. Chromatogr.*, *124*, 111 (1976).

19. E. G. G. Lindeberg, *J. Chromatogr.*, *117*, 439 (1976).

20. M. Roth, *Anal. Chem.*, *43*, 880 (1971).

21. H. Nakamura, *J. Chromatogr.*, *131*, 215 (1977).

22. N. C. Jain and R. H. Cravey, *J. Chromatogr. Sci.*, *12*, 228 (1974).

23. R. H. Cravey and N. C. Jain, *J. Chromatogr. Sci.*, *12*, 237 (1974).

24. S. J. Mule, *J. Chromatogr. Sci.*, *12*, 245 (1974).

25. M. L. Bastos and D. B. Hoffman, *J. Chromatogr. Sci.*, *12,* 269 (1974).

26. J. H. Speaker, *J. Chromatogr. Sci.*, *12,* 297 (1974).

27. J. A. F. de Silva, I. Berkersky, and C. V. Publisi, *J. Pharm. Sci.*, *63*, 1837 (1974).

28. D. B. Farber, C. Mulder, and W. A. Man In 'T Veld, *J. Chromatogr.*, *100*, 55 (1974).

29. D. B. Farber and W. A. Man In 'T Veld, *J. Chromatogr.*, *93*, 238 (1974).

30. W. S. Simmons and R. L. DeAngelis, *Anal. Chem.*, *45*, 1538 (1973).

31. F. Van Hoof and A. Heyndrickx, *Anal. Chem.*, *46*, 286 (1974).

32. C. W. Sigel, J. L. Woolley, Jr., and C. A. Nichol, *J. Pharm. Sci.*, *64*, 973 (1975).

33. F. Abe and K. Samejima, *Anal. Biochem.*, *67*, 298 (1975).

34. J. H. Fleisher and D. H. Russell, *J. Chromatogr.*, *110*, 335 (1975).

35. O. Heby and G. Andersson, *J. Chromatogr.*, *145*, 73 (1978).

36. D. B. Farber, A. De Kok, and U. A. Th. Brinkman, *J. Chromatogr.*, *143*, 95 (1977).

37. R. L. DeAngelis, M. F. Kearney, and R. W. Welch, *J. Pharm. Sci.*, *66*, 841 (1977).

38. J. L. Straughan, W. F. Cathcart-Rake, D. W. Shoeman, and D. L. Azarnoff, *J. Chromatogr.*, *146*, 473 (1978).

39. R. N. Gupta, F. Eng, and D. Lewis, *Anal. Chem.*, *50*, 197 (1978).

40. S. L. Tripp, E. Williams, W. J. Roth, W. E. Wagner, Jr., and G. Lukas, *Anal. Lett.*, *B11*, 727 (1978).

41. I. R. Kupke and S. Zeugner, *J. Chromatogr.*, *146*, 261 (1978).

42. P. J. van der Merwe and J. M. Steyn, *J. Chromatogr.*, *148*, 549 (1978).

43. R. N. Gupta, F. Eng, D. Lewis, and C. Kumana, *Anal. Chem.*, *51*, 455 (1979).

44. M. Schafer and E. Mutschler, *J. Chromatogr.*, *169*, 477 (1979).

45. A. Szent-Gyorgyi, *Science*, *126*, 751 (1957).

46. M. P. Gordon and D. South, *J. Chromatogr.*, *10*, 513 (1963).

47. K. Randerath, *Anal. Biochem.*, *21*, 480 (1967).

48. R. T. Mayer, G. M. Holman, and A. C. Bridges, *J. Chromatogr.*, *90*, 390 (1974).

49. I. Isenberg, M. J. Smerdon, J. Cardenas, J. Miller, H. W. Schaup, and J. Bruce, *Anal. Biochej.*, *69*, 531 (1975).

50. C. M. O'Donnell and J. D. Winefordner, *Clin. Chem.*, *21*, 285 (1975).

51. J. A. F. de Silva, N. Strojny, and K. Stika, *Anal. Chem.*, *48*, 144 (1976).

52. H. K. Y. Lau and G. G. Guilbault, *Enzym. Tech. Dig.*, *3*, 164 (1974).

53. G. G. Guilbault, "Analytical Uses of Immobilized Enzymes," in *Insolubilized Enzymes*, M. Salmona, C. Saronio, and S. Garattini, eds., Raven Press, New York, 1974.

54. G. G. Guilbault, "Fluorescence Analysis on Solid Surfaces," in *Analytical Chemistry: Essay in Memory of Anders Ringbom*, E. Wanninen, ed., Pergamon Press, New York, 1977.

55. G. G. Guilbault, *Photochem. Photobiol.*, *25*, 403 (1977).

56. G. G. Guilbault and R. L. Zimmerman, Jr., *Anal. Lett.*, *3*, 133 (1970).

57. G. G. Guilbault and A. Vaughan, *Anal. Chim. Acta*, *55*, 107 (1971).

58. R. L. Zimmerman, Jr. and G. G. Guilbault, *Anal. Chim. Acta*, *58*, 75 (1972).

59. H. K. Y. Lau and G. G. Guilbault, *Clin. Chem.*, *19*, 1045 (1973).

60. J. W. Kuan, H. K. Y. Lau, and G. G. Guilbault, *Clin. Chem.*,
     21, 67 (1975).

61. B. Rietz and G. G. Guilbault, *Clin. Chem.*, *21*, 715 (1975).

62. B. Rietz and G. G. Guilbault, *Clin. Chem.*, *21*, 1791 (1975).

63. S. W. Kiang, J. W. Kuan, and G. G. Guilbault, *Clin. Chem.*,
     21, 1799 (1975).

64. A. R. Sperling, *J. Chromatogr. Sci.*, *12*, 265 (1974).

65. R. L. Ranieri and J. L. McLaughlin, *J. Chromatogr.*, *111*, 234
     (1975).

66. M. Prosek, E. Kucan, M. Katic, and M. Bano, *Chromatographia*,
     9, 273 (1976).

67. M. Prosek, E. Kucan, M. Katic, and M. Bano, *Chromatographia*,
     10, 147 (1977).

68. E. M. Karlsson and H. W. Peter, *J. Chromatogr.*, *155*, 218 (1978).

69. M. I. Ichimasa, Y. Ichimasa, and K. Uranaka, *Agric. Biol. Chem.*,
     40, 1253 (1976).

70. J. A. Vinson, J. E. Hooyman, *J. Chromatogr.*, *135*, 226 (1977).

71. J. Sherma, M. F. Dobbins, and J. C. Touchstone, *J. Chromatogr.
     Sci.*, *12*, 300 (1974).

72. R. C. Thapliyal and J. L. Maddocks, *J. Chromatogr.*, *160*, 239
     (1978).

73. P. J. Cashion, H. J. Notman, T. B. Cadger, and G. M. Sathe,
     Anal. Biochem., *80*, 654 (1977).

74. T. Nakai, T. Ohta, N. Wanaka, and D. Beppu, *J. Chromatogr.*,
     88, 356 (1974).

75. J. J. Sabatka, D. A. Brent, J. Murphy, J. Charles, J. Vance,
     and M. H. Gault, *J. Chromatogr.*, *125*, 523 (1976).

76. W. A. Taylor, K. G. Blass, and C. S. Ho, *J. Chromatogr.*, *168*,
     501 (1979).

13

Other Applications

In this chapter, applications that did not fit readily into the previous chapters will be considered in greater detail.

GENERAL APPLICATIONS

Young (1) developed a fluorescence method for detecting aldehydes at the picomole level on silica gel chromatoplates. In the procedure, standard solutions containing 1 µl or less of aldehydes and ketones were spotted on a chromatoplate and then overspotted with 5 µg of aniline. The chromatoplate was developed, irradiated for 5 min with ultraviolet radiation, and then sprayed sequentially with fluorescamine and triethanolamine, and finally, viewed under ultraviolet radiation. The basis for the method was arrived at after observing chromatoplates on which a mixture of aniline and benazldehyde spontaneously formed the corresponding imine. Also, upon ultraviolet irradiation the imine regenerated aniline, and a yellow fluorescence resulted after treatment with fluorescamine and viewing under ultraviolet radiation.

$$R\text{-}CHO + H_2N\text{-}C_6H_5 \underset{UV}{\overset{TLC}{\rightleftarrows}} R\text{-}CH\text{=}N\text{-}C_6H_5$$

A variety of aldehydes were tested. The majority of aromatic aldehydes yielded imines and gave detection limits of 20-30 ng.

Those aldehydes having 4-hydroxy or 4-dimethylamino functionalites
yielded colored imines and did not give observable fluorescence.

Kiang, Kuan, and Guilbault (2) developed an enzymatic method
for the determination of nitrate in water. The enzyme NADH-dependent
nitrate reductase (EC 1.6.6.1) was induced from *Chlorella vulgaris*
and was purified by affinity chromatography. Nitrate was reduced
to nitrite by the above enzyme in the presence of NADH as electron
donor. The rate of disappearance of the fluorescence of NADH was
monitored by the silicon rubber pad fluorometric technique described
in Chapters 7 and 13. The reproducibility and accuracy of the method
were very good and the detection range was 50 ppb to 7.5 ppm. The
results from interferences are shown in Table 13.1, which indicate
that most of the anions and cations studied do not interfere. The
authors commented that the procedure could be a very useful method
for monitoring nitrate in contaminated drinking water, in foods, and
in other samples.

DeLeenheer, Sinsheimer, and Burckhalter (3) described an improved procedure for the determination of primary and secondary
amines based on the fluorescence of a cyclized product derived from
isothiourea derivatives of amines and 9-isothiocyanatoacridine.
Initial isothiourea formation and its cyclization were carried out
in toluene which resulted in the formation of highly fluorescent
2-alkylamino-1,3-thiazino-[4,5,6-kℓ]acridine. The fluorescence
of the thiazinoacridine compounds was determined either directly
in solution or after the separation on silica gel chromatoplates.
Compounds studied were n-butylamine, cyclohexylamine, benzylamine,
β-phenylethylamine, amphetamine, diethylamine, and N-methylpiperazine.
Amphetamine was detected at the 3-pl level.

Sherma and Marzoni (4) reported the detection and quantitation
of aniline and 10 substituted derivatives on silica gel chromatoplates by spraying with fluorescamine reagent; they performed their
densitometric quantitation by in situ scanning of the fluorescent
spots with a Kontes densitometer. They chose those particular anilines
because the anilines were related to important amide, carbamate,

TABLE 13.1 Interference Studies of Diverse Substances in the Fluorometric Method[a]

Diverse ions	Final concentration of diverse ions (M)	Error observed (%)
Ca^{2+}	1×10^{-3}	+2.0
Zn^{2+}	1×10^{-3}	-3.9
Mn^{2+}	1×10^{-3}	-5.2
Cu^{2+}	1×10^{-3}	-89.3
Co^{2+}	1×10^{-3}	+1.3
Ni^{2+}	1×10^{-3}	-18.4
Mg^{2+}	1×10^{-3}	-2.1
Hg^{2+}	1×10^{-3}	-100.0
Na^{+}	1×10^{-3}	-0.8
K^{+}	1×10^{-3}	-1.7
NH_4^{+}	1×10^{-3}	+3.1
NO_2^{-}	1×10^{-3}	-12.5
ClO_4^{-}	1×10^{-3}	-2.3
SO_4^{2-}	1×10^{-3}	-1.4
SO_3^{2-}	1×10^{-3}	+3.2
Cl^{-}	1×10^{-3}	-1.5
ClO_3^{-}	1×10^{-4}	+94.4
BrO_3^{-}	1×10^{-4}	+34.2
IO_3^{-}	1×10^{-4}	+18.7

[a] Nitrate concentration, 8×10^{-5} M.

Reprinted with permission from C. Kiang, S. S. Kuan, and G. G. Guilbault, *Anal. Chem.*, *50*, 1323 (1978). Copyright by the American Chemical Society.

and urea pesticides. Table 13.2 shows the minimum visual detection limits for the anilines.

Segura and Gotto (5) developed a fluorometric procedure for the detection and quantitation of numerous organic compounds on

TABLE 13.2 Minimum Visual Detectable Levels of Anilines with Fluorescamine Reagent

Compound	Chromatographic solvent[a]	Level (ng)
Aniline	A	4
p-Bromoaniline	A	4
p-Chloroaniline	A	4
m-Chloroaniline	A	10
3,4-Dichloroaniline	A	10
3-Chloro-4-methylaniline	A	10
m-Trifluoromethylaniline	A	10
3-Chloro-4-bromoaniline	A	10
m-Aminophenol	B	30
2-Amino-4-chlorophenol	B	30
p-Aminophenol	B	80

[a] Solvent A was benzene/ethanol (95:5 v/v); solvent B was benzene/methanol (80:20 v/v).

Reprinted with permission from *American Laboratory*, Volume 6, p. 26, 1974. Copyright 1974 by International Scientific Communications, Inc.

thin-layer chromatoplates. The formation of fluorescent derivatives was induced in most experiments by thermal treatment of the chromatoplates in the presence of ammonium hydrogen carbonate. In some experiments ammonium hydroxide was used, and in a few experiments the effect of heating alone and the effect in the presence of nitrogen and carbon dioxide were studied. Generally, chromatoplates were placed inside a sealed tank containing 6 g of ammonium hydrogen carbonate. The prepared tank was placed inside an oven and heated between 110 and 150°C for variable periods of time (2-12 hr), depending on the type of adsorbent and the nature of the compounds. It appeared aluminia was more effective than silica gel in inducing the formation of fluorescent derivatives. In situ fluorescence measurements were made with a Farrand MK-1 spectrofluorometer and

attachment for thin-layer chromatography (TLC) scanning. Excitation
and fluorescence maxima were given for several lipids, steroids,
sugars, amino acids, amino acid derivatives, purine and pyrimidine
derivatives, and miscellaneous compounds. Calibration curves were
given for some compounds. For lipids, steroids, and sugars, 0.1
µg could be detected easily. Catecholamines could be quantitated
at the low nanogram level.

Curtis and Seitz (6) described a new method for detecting fluorescent compounds by chemiluminescence. Dansyl derivatives of three
amino acids were separated on silica gel chromatoplates and were
sprayed successively with solutions of bis-2,4,6-trichlorophenyl-
oxalate (TCPO) and hydrogen peroxide in dioxane. The TCPO and peroxide
reacted to produce a dioxetane-dione intermediate which transferred
its energy to the dansyl compound causing it to luminesce. The
potential advantages of using chemiluminescence instead of fluorescence detection for TLC are these: no need for a light source, no
error from scattered radiation, and fewer instrumental geometry
problems. The authors considered solvents for TCPO and hydrogen
peroxide, effect of TCPO concentration and effect of peroxide concentration. They demonstrated that it was possible to observe the
fluorescent derivative of glycine on a silica gel chromatoplate
by spraying with TCPO and hydrogen peroxide.

Uchiyama and Uchiyama (7) discussed the fluorescence enhancement in TLC by spraying with viscous organic solvents. They investigated the physicochemical factors of the reagents that affect fluorescence enhancement using 5-dimethylamino-1-naphthalenesulfonamides
(DANS-amines) as test compounds. Nonpolar, viscous and nonacidic
solvents such as a mixture of liquid paraffin and n-hexane were
appropriate reagents for the DANS-amines, which enhanced the fluorescence intensity tenfold. With benzo[a]pyrene, a 35-fold enhancement
of fluorescence was observed. Most of the work was done with silica
gel chromatoplates, but fluorescence enhancement with DANS-dimethyl-
amine and benzo[a]pyrene were also observed on other adsorbents
such as polyamides and cellulose.

INORGANIC SPECIES AND METAL CHELATES

A fluorescent ring-oven technique for the microdetermination of
lead was reported by Skuric et al. (8). A piece of filter paper
was placed on a Weisz ring-oven maintained at 105-110°C. Portions
of sample solution were placed at the center of the paper and dried.
Then 0.05 M sulfuric acid was added to obtain the sulphate of the
cations present. Next the soluble sulphates were washed to the
periphery with several portions of 10^{-4} M sulfuric acid until the
whole paper was wet. No losses of lead occurred down to 0.05 μg
of lead with 10^{-4} M sulfuric acid. The paper was dried and lead
sulphate was washed to the ring zone with several portions of 0.1
M sodium chloride solution, thus obtaining fluorescent rings of
sodium tetrachloroplumbate. The fluorescence intensity of the unknown ring was visually matched with the fluorescence of a set of
standard lead rings under ultraviolet radiation. The sensitivity
was 0.05 μg, the analytical range was 0.1-5 μg, and the relative
standard deviation was about 20-30%. Twenty-nine foreign ions were
tested as potential interfering components. Permanganate exerted
a positive interference at 10 times the concentration of lead.
Barium and bismuth ions interfered to some extent with the quantitative washing out of the lead chlorocomplex into the ring.

Turina (9) reported a method in which small amounts of lead
could be detected on cellulose chromatoplates as the fluorescent
$PbCl_4^{2-}$ complex. The approach was used to determine lead in used
lubricating oil. The cellulose chromatoplate was developed to 2
cm with acetone:10 N hydrochloric acid:water (7:1:2), and then
dried. The sample was converted into the desired fluorescent complex during this process. Development was continued with the same
solvent until the mobile phase migrated to 10 cm from the origin.
The developed chromatoplate was thoroughly dried and then the fluorescent spots detected with an ultraviolet hand lamp. Reflected fluorescence was measured quantitatively with an Opton spectrophotometer
with a scanning accessory. Amounts as small as 0.2-0.3 μg of lead
per spot could be detected, and the detection limit was slightly
higher with paper chromatograms.

Ryan, Holzbecher, and Rollier (10) described the determination
of traces of lead(II) by solid state luminescence. Lead was coprecipitated with calcium oxalate, ignited to CaO:Pb phosphor, and
the luminescence was measured at 530 nm. The sample cell was a blackened aluminum sheet with drilled holes, covered with a thin glass plate.
The holes were packed with sample powder. The fluorescence intensity
was a linear function of lead concentration from 0.5 to 200.0 µg,
and the relative standard deviation of six determinations of 0.1
µg/l lead was approximately 9%. Lead was determined successfully
in the presence of a 100-fold excess of Al(III), Cr(III), Mg(II),
Hg(II) or Zn(II). A tenfold excess of Fe(III) and equivalent amounts
of Cd(II), Cu(II) and Ag(I) gave no interference. Co(II), Mn(II),
and Ni(II) at the 10-µg level interfered seriously. The only element
tested to give fluorescence was bismuth(III), but bismuth could be
tolerated at a tenfold excess. The method was applied successfully
to the direct determination of lead in leaves and in synthetic blood.
Lead could also be determined in samples containing interfering
components after separation by solvent extraction.

Rollier and Ryan (11) considered the solid state luminescence
of chelates in trace metal analysis using the aluminum-oxine system.
They were interested in determining whether preconcentration of
a microconstituent by using a carrier precipitate of organic reagent
would improve the sensitivity of the fluorescence determination
of metal chelates. Aspects such as the pH of precipitation and
the amount of reagent were investigated. Generally the procedure
involved slow precipitation of aluminum in a buffered solution with
oxine, then filtering and drying the precipitate. The dried precipitate was ground and mixed, and packed in a cell, and the luminescence was measured at 524 nm. The cell they used was described above.
The calibration curve was linear from 0.1 to 10 µg. The relative
standard deviation of six determinations of 10 µg of aluminum was
2.2% for 10-ml samples and 4.8% for 100-ml samples. Interference
studies showed severe interference from Cu, Zn, Fe, Mn, and Ca.
The coprecipitation technique did not improve the selectivity of

the determination. Masking and extraction techniques were investigated as ways to eliminate interferences. With cyanide as a masking
agent, copper did not interfere, but Zn, Fe and Mn did interfere.
Extraction procedures were not successful in preventing interference
problems. The authors stated that a good reagent for solid state
luminescence should be nonfluorescent, insoluble, and selective.
A reagent that has these qualities is 1,3-di(2-naphthyl)-1,3-propane-
dione.

Delumyea and Schenk (12) studied lead(II)-manganese energy
transfer in sodium chloride pellets from a quantitative viewpoint.
Lead(II) and manganese(II) were coprecipitated from a saturated
sodium chloride solution by adding ethanol. After filtration and
drying, the sodium chloride matrix was compressed by standard pellet
techniques into a crystalline pellet. The excitation maxima at 275
and 303 nm were obtained for lead(II) and a broad emission maxima
at 610 nm was obtained for manganese(II). The coprecipitation of
lead(II) in an excess of manganese(II) was adequate to yield a reproducible analytical luminescence curve. A special pellet holder
was constructed for a Turner filter fluorometer to permit routine
fluorometric measurements of traces of lead(II). Table 13.3 shows
that the percentage lead coprecipitated was generally reproducible
over a wide range of concentrations.

Table 13.4 gives results for interference by metal ions on a
standard lead(II) sample. The cations iron(III), mercury(II), and
tin(II) posed potentially serious interference problems. The authors
emphasized that with sufficient study multielement analysis could
be performed on a single pellet. They listed several systems exhibit-
ing sensitized luminescence.

Miller et al. (13) developed a spectroscopic system for the
study of fluorescent lanthanide probe ions in solids. Lanthanide
ions can be used as fluorescent probes for the study of defects
in solids and for qualitative and quantitative determination of
trace concentrations of foreign ions included in the lattices.
They described a spectroscopic system based on a tuneable dye laser

TABLE 13.3 Study of Reproducibility of Coprecipitation of Lead(II) in Sodium Chloride

Lead(II) added (μg)[a]	Lead(II) coprecipitated (μg)[b]	Recovery (%)
10	9.89	98.9
20	21.54	107.7
50	49.72	99.4
100	74.12	74.1
200	144.8	72.4
500	485.9	97.2
1000	969.8	97.0
		Average 92.4 ± 10.8

[a] Amount calculated to be in 0.50 g of sodium chloride precipitate.

[b] Actual quantity observed in 0.50-g sample of precipitate upon atomic absorption analysis.

Reprinted with permission from R. G. Delumyea and G. H. Schenk, *Anal. Chem.*, *48*, 95 (1976). Copyright by the American Chemical Society.

TABLE 13.4 Interference by Metal Ions on a Standard Lead(II) Sample[a]

Metal ion	0 ppm	4 ppm	8 ppm	20 ppm	40 ppm
Iron(III)	66	56	42	25	25
Mercury(II)	58	56	45	37	23
Copper(II)	39	44	40	40	48
Magnesium(II)	55	56	57	46	45
Zinc(II)	56	61	51	58	--
Cobalt(II)	62	69	50	55	--
Tin(II)	30	60	∼80	--	--

[a] Luminescence of sample containing metal ion. Data taken on a Turner filter fluorometer.

Reprinted with permission from R. G. Delumyea and G. H. Schenk, *Anal. Chem.*, *48*, 95 (1976). Copyright by the American Chemical Society.

excitation source capable of selectively exciting the probe ions
present at different crystallographic sites. Techniques for measuring fluorescent lifetimes using signal averaging were developed,
and a computer controlled system for obtaining excitation and absorption spectra simultaneously was also developed. The system allowed
the determination of radiative quantum efficiencies and the calcula-
tion of absolute site concentrations.

SENSITIZED FLUORESCENCE

Energy transfer between an excited singlet donor molecule and a
ground singlet acceptor molecule to produce an excited acceptor
molecule which undergoes radiative deactivation is called *sensitized
fluorescence* (14). Hornyak (15) described the determination of
naphthacene by sensitized fluorescence on filter paper. Naphthacene
is weakly fluorescent with a quantum efficiency of 0.002. When
trace quantities of naphthacene are built into the crystal lattice
of another polycyclic aromatic hydrocarbon such as 2,3-benzfluorene
and excited at an excitation wavelength of the energy donor, the
characteristic yellow-green sensitized fluorescence of naphthacene
appears. Some of the adsorbed donor energy is transferred by radia-
tionless transition to naphthacene molecules. The naphthacene decreases strongly the fluorescence of the energy donor, and this
change is proportional to the concentration of naphthacene. Hornyak
dissolved 2,3-benzfluorene and naphthacene in benzene to give solu-
tions which were 5×10^{-5} M, and 10^{-8} to 5×10^{-5} M, respectively.
Whatman paper strips were immersed in the above solutions, and after
10 min the solvent was evaporated, and fluorescence intensities
of both 2,3-benzfluorene and naphthacene were determined. The fluorescence intensities were measured at 491 nm for naphthacene and 396
nm for 2,3-benzfluorene. The ratio of these intensities was plotted
against the naphthacene concentration, and the curve was linear from
3×10^{-7} M to 5×10^{-5} M naphthacene. A standard deviation of 1-2%
was achieved.

Hornyak (16) further considered the sensitized fluorescence of
naphthacene by benzo[a]pyrene and its analytical application. The
analytical procedure was similar to that described in the preceding
paragraph. Benzo[a]pyrene can sensitize the emission of other polycyclic aromatic compounds fluorescing weakly or not at all, if incident radiation is used corresponding to the absorption band of
benzo[a]pyrene. With naphthacene, as a result of energy transfer,
the fluorescence of benzo[a]pyrene decreases substantially, while
the characteristic emission of naphthacene becomes dominant. Hornyak
plotted the ratio of the fluorescence intensities of naphthacene
(498 nm) to benzo[a]pyrene (456 nm) against naphthacene concentration.
The resulting curve was suitable for analytical purposes, and concentrations as little as 3×10^{-9} M naphthacene could be determined.

Hornyak (17) considered the sensitized fluorescence of naphthacene by 1,2-benzanthacene on Whatman paper. As with 2,3-benz-
fluorene and benzo[a]pyrene, naphthacene considerably reduced the
fluorescence of 1,2-benzanthracene. The calibration curve was linear
from 10^{-7} to 10^{-4} M naphthacene, and the relative standard deviation
was lower than 1%.

POLYMERS

The direct determination of inhibitors in polymers by luminescence
techniques was considered by Drushel and Sommers (18). Inhibitors
are added to polymers to prevent oxidation and degradation, which
cause undesirable changes in polymer physical properties. Several
common inhibitors were found to have strong luminescence for possible quantitative use. Drushel and Sommers developed procedures
for the direct determination of some selected inhibitors in polymer
films. Both room temperature and liquid-nitrogen temperature were
used in the luminescence measurements. Age Rite D, a polymeric
form of trimethyldihydroquinoline, exhibited an intense fluorescence.
For the measurement of Age Rite D directly in pressed polymer films,
it was necessary to keep the films less than 0.01 cm to prevent

concentration quenching. Concentration quenching and background absorption were also considered in relation to their effect on the linearity of calibration curves and the precision of the luminescence measurements. Quantum efficiencies for the absorption-emission processes were also determined.

Allen, Homer, and McKellar (19) reported the fluorescence and phosphorescence properties of a variety of commercial polymers. The results suggested that luminescence analysis should be very useful in the characterization of a wide range of polymers. A Hitachi Perkin-Elmer MPF-4 spectrofluorometer with phosphorescence attachments was used to obtain the luminescence data. Polymer samples in the form of granules and fibers were examined for fluorescence in a silica Dewar flask that contained no nitrogen. Powder samples were placed in 5-mm o.d. silica tubes before positioning the tubes in the Dewar flask. Phosphorescence measurements on granules, fibers, and powder materials were performed as above except liquid nitrogen was in the Dewar flask and the phosphoroscope was in position. Phosphorescence lifetimes were measured with the aid of a Tetronix DM-64 storage oscilloscope. The authors measured the following properties in the characterization of commercial polymers: fluorescence emission and excitation spectra, phosphorescence emission and excitation spectra, and phosphorescence lifetime. The luminescence characteristics were given for 18 different polymers. The fluorescence and phosphorescence excitation wavelength maxima for many polymers were different, indicating that the same light-absorbing species were not responsible for both emissions. Most of the polymers showed marked differences in their phosphorescence lifetimes. In some cases, a change in the physical form of the polymer altered the phosphorescence lifetime. Possibly, differences in lifetime were due to changes that occurred in the morphological structure of the polymer. The advantages of luminescence spectroscopy for characterizing polymers are as follows: (1) the approach is nondestructive, rapid, and sensitive; (2) no tedious sample preparation is required; (3) powder, chip, film, or fiber form can be analyzed; (4) polymers

can be characterized within a particular class such as polyamides.
The disadvantages are as follows: (1) the technique gives no information on commercial polymers that are not luminescent; (2) some light
stabilizers quench the luminescence of polymers.

Luminescence spectroscopy has been used considerably in theoretical and structural studies of polymers. Nishijima (20) reviewed
fluorescence methods in polymer research. He considered such aspects as polarization characteristics of fluorescence and molecular
orientation patterns. Morawetz (21) discussed some applications
of fluorometry to synthetic polymer studies. He reported on fluorescence quenching, depolarization of fluorescence, excimer fluorescence, and nonradiative energy transfer. Allen et al. (22) described
a study of the phosphorescence emission from the commercial poly-
olefins, i.e., polyethylene, polypropylene, and poly-4-methylpentene-1,
and the polyamides, nylon 6/6, 6, 11, and 12. Impurity centers in
the polymers were responsible for the emission and were present
either as groups directly attached to the polymer backbone or as
discrete molecules within the polymer matrix. The authors were
interested in the part the impurities played in the thermal and
photochemical oxidation of the polymers. Changes in the phosphorescence spectra provided valuable information on the mechanisms of both
degradation processes.

REFERENCES

1. J. C. Young, *J. Chromatogr.*, *130*, 392 (1977).

2. C. Kiang, S. S. Kuan, and G. G. Guilbault, *Anal. Chem.*, *50*,
   1323 (1978).

3. A. DeLeenheer, J. E. Sinsheimer, and J. H. Burckhalter, *J.
   Pharm. Sci.*, *62*, 1370 (1973).

4. J. Sherma and G. Marzoni, *Amer. Lab.*, *6*, 21 (1974).

5. R. Segura and A. M. Gotto, Jr., *J. Chromatogr.*, *99*, 643 (1974).

6. T. G. Curtis and W. R. Seitz, *J. Chromatogr.*, *134*, 343 (1977).

7. S. Uchiyama and M. Uchiyama, *J. Chromatogr.*, *153*, 135 (1978).

8. Z. Skuric, F. Vlic, and J. Prpic-Marecic, *Anal. Chim. Acta*,
   73, 213 (1974).

9. N. Turina, *J. Chromatogr., 93*, 211 (1974).

10. D. E. Ryan, J. Holzbecher, and H. Rollier, *Anal. Chim. Acta, 73*, 49 (1974).

11. H. Rollier and D. E. Ryan, *Anal. Chim. Acta, 74*, 23 (1975).

12. R. G. Delumyea and G. H. Schenk, *Anal. Chem., 48*, 95 (1976).

13. M. P. Miller, D. R. Tallant, F. J. Gustafson, and J. C. Wright, *Anal. Chem., 49*, 1474 (1977).

14. J. D. Winefordner, S. G. Shulman, and T. C. O'Haver, *Luminescence Spectrometry in Analytical Chemistry*, Wiley-Interscience, New York, 1972, p. 311.

15. I. Hornyak, *Anal. Chim. Acta, 80*, 393 (1975).

16. I. Hornyak, *Acta Chim. Acad. Sci. Hung., Tomus, 94*, 87 (1977).

17. I. Hornyak, *Mikrochim. Acta*, 23 (1978).

18. H. V. Drushel and A. L. Sommers, *Anal. Chem., 36*, 836 (1964).

19. N. S. Allen, J. Homer, and J. F. McKellar, *Analyst, 101*, 260 (1976).

20. Y. Nishijima, *Prog. Polym. Sci. (Japan), 6*, 199 (1973).

21. H. Morawetz, *Science, 203*, 405 (1979).

22. N. S. Allen, J. Homer, J. F. McKellar, and G. O. Phillips, *Br. Polym. J., 7*, 11 (1975).

14

Future Trends

Future research should center around theory, instrumentation, and applications. Several comments and suggestions were made in the earlier chapters on work that could be done in these areas. In this chapter, many of those ideas will be summarized and additional suggestions made for techniques that could be pursued to make solid surface luminescence analysis even more versatile.

In Chapter 4 the status of the analytical theory of solid surface luminescence was discussed. As was emphasized, there is a strong need for theoretical equations that describe luminescence reflected or transmitted from solid surfaces that have been substantiated by experimental data. More research under a variety of experimental and instrumental conditions is needed to establish the validity of recent theoretical equations. Such aspects as layer thickness, particle size of solid materials, absorption of exciting radiation by the solid surfaces, intensity of luminescent signals in the reflected and transmitted modes, simultaneous measurement of reflected and transmitted luminescent radiation, and angle of observation of the luminescent signal should be considered. With established theoretical equations, the analyst can predict such things as linearity of calibration curves and conditions for optimum sensitivity.

Recent work with room temperature phosphorescence has shown that several solid surfaces can be employed to obtain useful analytical

data. However, much research can be done in the future to find
other surfaces that will induce room temperature phosphorescence
from organic compounds. For example, a much better understanding
of the molecular interactions between the surface and the phosphorescent molecules is needed. Some work in this direction has been initiated (1-4).

Commercial instruments and instruments designed by researchers
were detailed in Chapters 2 and 3. Generally, commercial and research
instruments are adequate for obtaining good luminescence data.
However, several improvements can be made. Positioning of the adsorbed component in the beam of source radiation does cause difficulty
and can cause errors that exceed 2% (5). More extensive use of
computers and digital electronics will minimize this source of error.
Ebel and Hocke (5,7,8) and Ebel, Herold, and Hocke (6) have discussed
some of these aspects already. In one application they employed
a programmable desk calculator with a spectrodensitometer to maximize the position of each adsorbed component with respect to the
source beam (5). The use of laser sources should fine greater application in the future for the reasons mentioned earlier in Chapter
3. Many of the new instrument principles and use of lasers that
have appeared recently for solution fluorescence should be applicable
to solid surface analysis (9-12). Because highly scattering media
usually are employed as solid surfaces, new detection systems can
be employed to improve signal to noise ratio. Television-type multi-
channel detectors (image devices) should find use in this area (13-19).
Cooney et al. (18) have compared image devices and photomultiplier
detectors in atomic and molecular luminescence spectrometry by signal-
to-noise ratio calculations. Their results should be particularly
helpful in designing new detection systems for solid surface luminescence analysis. In addition, derivative spectroscopy and simultaneous multicomponent analysis should be possible (15). Also, photon
counting systems should find application in solid surface luminescence analysis (20,21). Finally, recent work with contour plotting
of fluorescence data from mixtures and three-dimensional plotting

of fluorescence data should be useful in solid surface luminescence
analysis (22-25).

Because of the speed, simplicity, sensitivity, and relatively
low cost of solid surface luminescence analysis, many applications
will appear in the future. The numerous applications presented in
earlier chapters support strongly the previous statement. The use
of luminescence detection in conjunction with both normal thin-layer
chromatography (TLC) and high-performance thin-layer chromatography
(HPTLC) should find wider application. This should be especially
true for HPTLC. Ripphahn and Halpaap (26) and Zlatkis and Kaiser
(27) have reviewed recently many aspects of HPTLC. Very recently
Kaiser (28) discussed the accuracy and precision in today's instru-
mentalized TLC in quantitative and qualitative analysis. He stated
that thin-layer chromatoplates of highly improved quality are now
available, and their chromatographic characteristics have come close
to those of HPLC columns. Reproducibility of repetitive scanning
of a single spot can be obtained at the 0.1% level. The reproducibility with the same sample and repeated separation in the nanogram
range was at the 1% level. Components separated by HPTLC can be
detected by fluorescence, and room temperature phosphorescence should
be applied successfully in the future. Also, it is easy to collect
a sample from an HPLC column, spot the sample on a solid surface,
and obtain luminescence data. Both fluorescence and room temperature phosphorescence data can be obtained.

In the future, solid surface luminescence analysis should be
thought of in general terms and considered as another dimension
in chemical analysis and not as just a technique used with TLC.
Analysts should use solid surfaces as they would use solutions for
chemical analysis. Small samples are easily handled by solid surface luminescence techniques, and chemical reactions can be carried
out readily on solid surfaces. Also, many times chemical changes
are more readily observable on solid surfaces than in solution.

REFERENCES

1. E. M. Schulman and R. T. Parker, *J. Phys. Chem.*, *81*, 1932 (1977).

2. R. M. A. von Wandruszka and R. J. Hurtubise, *Anal. Chem.*, *49*, 2164 (1977).

3. C. D. Ford and R. J. Hurtubise, *Anal. Chem.*, *52*, 656 (1980).

4. E. L. Yen-Bower and J. D. Winefordner, *Anal. Chim. Acta, 102,*

5. S. Ebel and J. Hocke, *J. Chromatogr.*, *126*, 449 (1976).

6. S. Ebel, G. Herold, and J. Hocke, *Chromatographia, 8*, 573 (1975).

7. S. Ebel and J. Hocke, *Chromatographia, 9*, 78 (1976).

8. S. Ebel and J. Hocke, *Chromatographia, 10*, 123 (1977).

9. D. C. Harrington and H. V. Malmstadt, *Anal. Chem.*, *47*, 271 (1975).

10. T. F. Van Geel and J. D. Winefordner, *Anal. Chem.*, *48*, 335 (1976).

11. J. H. Richardson and S. M. George, *Anal. Chem.*, *50*, 616 (1978).

12. R. M. Wilson and T. L. Miller, *Anal. Chem.*, *47*, 256 (1975).

13. Y. Talmi, *Anal. Chem.*, *47*, 658A (1975).

14. Y. Talmi, *Anal. Chem.*, *47*, 697A (1975).

15. Y. Talmi, D. C. Baker, J. R. Jadamec, and W. A. Saner, *Anal. Chem.*, *50*, 936A (1978).

16. J. R. Jadamec, W. A. Saner, and Y. Talmi, *Anal. Chem.*, *49*, 1316 (1977).

17. R. P. Cooney, T. Vo-Dinh, and J. D. Winefordner, *Anal. Chim. Acta, 89*, 9 (1977).

18. R. P. Cooney, T. Vo-Dinh, G. Walden, and J. D. Winefordner, *Anal. Chem.*, *49*, 939 (1977).

19. R. P. Cooney, G. D. Boutilier, and J. D. Winefordner, *Anal. Chem.*, *49*, 1048 (1977).

20. J. D. Winefordner, S. G. Schulman, and T. C. O'Haver, *Luminescence Spectrometry in Analytical Chemistry*, Wiley-Interscience, New York, 1972, p. 182.

21. V. Pollak, *J. Chromatogr.*, *133*, 49 (1977).

22. I. M. Warner, J. G. Callis, E. R. Davidson, M. Gouterman, and G. D. Christian, *Anal. Letters, 8*, 665 (1975).

23. I. M. Warner, G. D. Christian, E. R. Davidson, and J. B. Callis, *Anal. Chem.*, *49*, 564 (1977).

24. D. W. Johnson, J. B. Callis, and G. D. Christian, *Anal. Chem.*, *49*, 747A (1977).

25. J. H. Rho and J. L. Stuart, *Anal. Chem.*, *50*, 620 (1978).

26. J. Ripphahn and H. Halpaap, *J. Chromatogr.*, *112*, 81 (1975).

27. A. Zlatkis and R. E. Kaiser, *High Performance Thin-Layer Chromatography*, Vol. 9, Elsevier Scientific Publishing Co., Amsterdam, 1977.

28. R. E. Eaiser, "Abstracts of Papers," 177th National Meeting of the American Chemical Society, Honolulu, Hawaii, April, 1979, Abstract Anal. 15.

Author Index

Numbers in brackets are reference numbers and indicate that an author's work is referred to although his name is not cited in the text. Italic numbers give the page on which the complete reference is listed.

Subject Index